Marrow Protection

Progress in Experimental Tumor Research

Vol. 36

Series Editor   *Joseph R. Bertino*, New York, N.Y.

Basel · Freiburg · Paris · London · New York ·
New Delhi · Bangkok · Singapore · Tokyo · Sydney

Marrow Protection

Transduction of Hematopoietic Cells with Drug Resistance Genes

Volume Editor   *Joseph R. Bertino*, New York, N.Y.

36 figures, and 21 tables, 1999

 Basel · Freiburg · Paris · London · New York ·
New Delhi · Bangkok · Singapore · Tokyo · Sydney

Joseph R. Bertino
Molecular Pharmacology and Therapeutics
Sloan-Kettering Institute for Cancer Research
1275 York Avenue
New York, NY 10021 (USA)

Library of Congress Cataloging-in-Publication Data
Marrow protection. Transduction of hematopoietic cells with drug resistance genes /
volume editor, Joseph R. Bertino
(Progress in experimental tumor research; vol. 36)
Includes bibliographical references and indexes.
1. Cancer – Chemotherapy – Complications – Gene therapy. 2. Hematopoietic stem cells.
3. Drug resistance in cancer cells. 4. Transduction. 5. Genetic transformation.
I. Bertino, Joseph R.   II. Series.
[DNLM: 1. Hematopoietic Stem Cells. 2. Drug Resistance, Neoplasm – Genetics. 3. Gene Therapy.
4. Gene Transfer.   W1 PR668T v.36 1999]
RC271.C5T69 1999
616.99'4061–dc21
ISSN 0079–6263
ISBN 3–8055–6828–2 (hardcover: alk. paper)

Bibliographic Indices. This publication is listed in bibliographic services, including Current Contents® and Index Medicus.

Drug Dosage. The authors and the publisher have exerted every effort to ensure that drug selection and dosage set forth in this text are in accord with current recommendations and practice at the time of publication. However, in view of ongoing research, changes in government regulations, and the constant flow of information relating to drug therapy and drug reactions, the reader is urged to check the package insert for each drug for any change in indications and dosage and for added warnings and precautions. This is particularly important when the recommended agent is a new and/or infrequently employed drug.

All rights reserved. No part of this publication may be translated into other languages, reproduced or utilized in any form or by any means electronic or mechanical, including photocopying, recording, microcopying, or by any information storage and retrieval system, without permission in writing from the publisher.

© Copyright 1999 by S. Karger AG, P.O. Box, CH–4009 Basel (Switzerland)
www.karger.com
Printed in Switzerland on acid-free paper by Reinhardt Druck, Basel
ISBN 3–8055–6828–2

Contents

VII Preface

1 Basic Principles of Gene Transfer in Hematopoietic Stem Cells
Sadelain, M.; May, C.; Rivella, S.; Glade Bender, J. (New York, N.Y.)

20 Optimizing Conditions for Gene Transfer into Human Hematopoietic Cells
Moore, M.A.S.; MacKenzie, K.L. (New York, N.Y.)

50 Transfer of the MDR-1 Gene into Hematopoietic Cells
Bank, A.; Ward, M.; Hesdorffer, C. (New York, N.Y.)

65 O^6-Benzylguanine-Resistant Mutant MGMT Genes Improve Hematopoietic Cell Tolerance to Alkylating Agents
Davis, B.M.; Koç, O.N.; Reese, J.S.; Gerson, S.L. (Cleveland, Ohio)

82 Use of Variants of Dihydrofolate Reductase in Gene Transfer to Produce Resistance to Methotrexate and Trimetrexate
Bertino, J.R.; Zhao, S.C.; Mineishi, S.; Ercikan-Abali, E.A.; Banerjee, D. (New York, N.Y.)

95 Augmentation of Methotrexate Resistance with Coexpression of Metabolically Related Genes
Mineishi, S. (Madison, Wisc.)

107 Protection of Bone Marrow Cells from Toxicity of Chemotherapeutic Agents Targeted toward Thymidylate Synthase by Transfer of Mutant Forms of Human Thymidylate Synthase cDNA
Banerjee, D.; Tong, Y.; Liu-Chen, X.; Capiaux, G.; Ercikan-Abali, E.A.; Takebe, N.; O'Connor, O.A.; Bertino, J.R. (New York, N.Y.)

115 Dihydropyrimidine Dehydrogenase and Resistance to 5-Fluorouracil
Diasio R.B. (Birmingham, Ala.)

124 Chemoprotection against Cytosine Nucleoside Analogs Using the Human Cytidine Deaminase Gene
Eliopoulos, N.; Beauséjour, C.; Momparler, R.L. (Montréal)

143 In vivo Selection of Hematopoietic Stem Cells Transduced with DHFR-Expressing Retroviral Vectors
Sorrentino, B.P.; Allay, J.A.; Blakley, R.L. (Memphis, Tenn.)

162 In vivo Selection of Genetically Modified Bone Marrow Cells
Blau, C.A. (Seattle, Wash.)

172 Gene Transfer in the Nonmyeloablated Host
Quesenberry, P.J.; Becker, P.S. (Worcester, Mass.)

179 Author Index

180 Subject Index

Preface

Successful transfer of drug resistance genes into hematopoietic stem cells affords great promise as a means of protecting patients from the myelotoxicity of anticancer drugs and may permit dose intensification and better outcomes. The use of peripheral blood mobilized progenitor cells has improved the access to this cell population and ex vivo gene transfer approaches. Transfer of drug resistance genes into peripheral blood mobilized progenitor cells together with nonselectable genes, also offers the possibility of expression of proteins produced by these genes and correction of genetic deficiencies or expression of other desirable proteins such as antibodies.

The studies published in this book review the progress in this field, emphasizing the work done with retroviral vectors and various genes that may be used to confer specific types of drug resistance to recipients. The requirement that progenitor cells have to go through mitosis to allow integration of the retrovirus has meant that stem cells must be pushed into cycle. Recent work with lentiviral constructs may obviate the need for cytokines to fulfil this purpose. Nevertheless, with improved retroviral constructs and titers, fibronectin and appropriate cytokine cocktails, and the possibility of in vivo selection, as discussed in this work, a new generation of clinical trials is on the horizon, which hopefully will lead to new insights and some success. It is hoped that this book will stimulate the pace of progress in this field.

Joseph R. Bertino, New York, N.Y.

Basic Principles of Gene Transfer in Hematopoietic Stem Cells

Michel Sadelain, Chad May, Stefano Rivella, Julia Glade Bender

Department of Human Genetics, Memorial Sloan-Kettering Cancer Center and Cornell University Medical College, New York, N.Y., USA

Gene transfer into hematopoietic stem cells (HSC) is one of the magna opera of gene therapy, representing the gateway to altering the genotype of blood cells and cells of the immune system. In order to achieve stable and prolonged effects, self-renewing HSC will have to be successfully transduced. Where transient therapeutic effects are sufficient or when repetitive therapy is acceptable, non-self-renewing progenitor cells would be suitable alternative target cells for gene transfer.

HSC capable of extensive self-renewal and pluripotent differentiation represent a minor population of adult bone marrow (<1 in 10^4–10^5 nucleated marrow cells). Their definition is functional: HSC are single cells able to repopulate the lymphohematopoietic system of lethally irradiated recipient mice and to give rise to progeny able to repopulate secondary recipients [1–3]. In recent years, better characterization of the HSC surface phenotype has improved their identification and enrichment. However, definitive agreement on what assays are valid surrogate measurements for bona fide HSC is still lacking at this time. Thus, the study of gene transfer in HSC remains intimately associated with that of HSC purification, proliferation, and differentiation.

The stable alteration of HSC and their progeny requires the use of vectors that either integrate in target cell chromosomes or stably replicate as extrachromosomal episomes. Whatever the vector, it is essential to assess whether the differentiation and self-renewing properties of the transduced cells are preserved. This chapter will review issues pertinent to the use of recombinant retroviruses as vectors. The focus will be on the biological principles that underly the construction of retroviral vectors for gene transfer in HSC and the regulation of gene expression in the progeny of transduced HSC.

Definition of Hematopoietic Stem Cells as Target Cells for Gene Transfer

The study of gene transfer and gene expression in HSC is complicated by the rarity of this cell population, the absence of valid surrogate cell lines, and the difficulty in assessing in each experiment whether one is investigating bona fide HSC or less primitive progenitor cells.

Human stem cell populations can express the CD34 antigen, as do many hematopoietic progenitors, certain endothelial populations and marrow fibroblast progenitors. A minor subset of human fetal liver and bone marrow CD34+ cells yield a progeny of myeloid, B cells, and thymocytes in SCID mice engrafted with human fetal thymic and bone marrow stroma [4]. The CD34+ HLA-DR− fraction of human marrow produces long-term lymphomyeloid lineage chimerism when transplanted in utero into immunoincompetent fetal sheep [5]. In the human autologous transplant setting, there is extensive literature documenting full hematopoietic reconstitution using CD34+-enriched peripheral blood mononuclear cells (PBMC) [6]. HSC enrichment typically involves positive selection for CD34 and depletion of cells with lineage markers, including CD38, which is present on most CD34+ hematopoietic cells and absent on stem cells [7]. Thus, no single marker is specific for stem cells, but combinations of markers can enrich HSC fractions by 100-fold or more. The fractions of bone marrow cells that exclude vital dyes such as rhodamine 123 [8] and Hoechst 33342 [9] are also enriched for cells with increased proliferative potential. In the studies of Goodell et al. [10], cells with HSC function were also identified in the CD34− fraction.

Primitive human hematopoietic cells are able to form multilineage 'cobblestone' areas on marrow stromal cells [11], a property which, in itself, does not represent a direct measurement of the HSC population. Long-term culture-initiating cells (LTC-IC) are identified by their ability to generate progenitor colonies after 5 weeks of culture on bone marrow stroma or immortalized hematopoietic stromal lines [12, 13]. Such cells comprise <1% of the CD34+ population in marrow, peripheral blood, or cord blood. More recently, it has been suggested that if long-term cultures are extended beyond 60 days, a functionally distinct and even smaller subpopulation of quiescent primitive cells with greater long-term generative capacity may be identified [14]. Thus, different surrogate in vitro assays are available to study cells with a large developmental potential, but it should not be forgotten that CD34+ cells are not equivalent to HSC.

Given the rarity of HSC and their predominantly quiescent state, it is a wonder that retroviral-mediated gene transfer could be achieved at all. However, the occurrence of gene transfer in stem cells has been obtained in a

Table 1. Factors affecting the efficiency of retroviral transduction

Viral components:	envelope (vector tropism)
	core (nuclear import, integration)
	viral titer (vector design, packaging cell line, virus concentration)
Target cell status:	proliferation (cytokines, stroma)
	receptor expression (cytokines, stroma, specific inducers)
	metabolism (ancillary support of reverse transcription)
Transduction conditions:	multiplicity of infection (target cell and virus concentration)
	adjuvants (polybrene, protamine sulfate, fibronectin)
	duration and repetition of exposure to virions
	physical conditions (flow, temperature, centrifugation)

number of murine bone marrow chimeras and formally established in cytokine-activated, cultured mouse bone marrow cells [15]. The challenge is to translate such occasional events into an efficient and reproducible process without diminishing the essential biological properties of stem cells.

Molecular and Cellular Requirements for Retroviral Gene Transfer in Hematopoietic Stem Cells

Retroviral-mediated gene transfer consists of an infectious process using a replication-defective retrovirus. Efficient infection requires the presence and activity of all molecules necessary for virion binding to the target cell, viral fusion, reverse-transcription, nuclear translocation of the preintegration complex, and proviral integration. The success of the procedure is dependent on optimized culture conditions, which have to balance the requirements for HSC expansion and preservation with those uniquely relevant to retroviral infection. A detailed analysis of all these factors is beyond the scope of this review. Key factors are listed in table 1. We will focus here on the components of retroviral vectors that are relevant to gene transfer in HSC.

Viral Envelope

Prior to fusion of the retroviral particle with the target cell cytoplasmic membrane, the viral envelope must engage specific cellular receptors. The envelope determines the tropism of a given family of retroviruses, which is restricted by the expression of the receptor on a given target cell [16]. The natural retroviral envelopes can be replaced by heterologous retroviral envelopes or alternative fusogenic molecules, a procedure referred to as pseudotyp-

Table 2. Packaging cell lines used for gene transfer in human cells

Packaging line	Parental cell line	Envelope	Reference
PA317	NIH 3T3	amphotropic	17
Ψ-CRIP	NIH 3T3	amphotropic	18
AM12	NIH 3T3	amphotropic	19
Propak	293	amphotropic	20
Phoenix	293T	amphotropic	21
PG13	NIH 3T3	GaLV	22
10A1	NIH 3T3	GaLV/ampho	28
gpg29	293	VSV-G	29

ing. Three viral envelopes have been used so far to target human cells: the amphotropic envelope of the murine leukemia virus (MLV strain 4070A) [17–21], the envelope of the gibbon ape leukemia virus (GaLV strain SEATO) [22], and the G glycoprotein of the vesicular stomatitis virus (VSV-G) [23]. These bind, respectively, to Ram-1 [24, 25], Glvr-1 [26], and phosphatidylserine and, possibly, additional cell-surface glycolipids [27]. A number of packaging cell lines are available to generate different pseudotypes of MLV-based retroviral vectors (table 2).

To date, gene transfer protocols utilizing the amphotropic retrovirus have shown disappointingly low transduction of HSC. In the mouse, where transduction efficiency can surpass the 10% range using ecotropic virus, infection with equal volume and similar viral titer of amphotropic virus yields a transduction efficiency that is <5% that of the ecotropic vector [30]. Even lower transduction efficiency was reported in long-term studies in other species, including dogs [31–33], primates [34, 35] and humans [36–39]. The poor transduction efficiency of both murine and human HSC by amphotropic virus could be in part caused by low levels of amphotropic receptor expression in these cells (of course, other factors contribute at the same time to limit transduction efficiency, as discussed below). Preselecting a population of HSC with elevated retrovirus receptor expression [30], inducing higher expression of the receptor by chemical stimulation [26] or by genetic means [40] increasing the viral multiplicity of infection, and increasing the duration of retroviral infection are logical steps to attempt to increase transduction efficiency [41].

Thus, although the receptor for amphotropic particles is widely distributed among many cell types, it remains to be established to what extent it is present on the surface of human HSC [30, 42–45]. Alternatives to the commonly used MLV amphotropic envelope may be better suited for the infection of HSC. This optimal fit may vary according to developmental stage, i.e. HSC of fetal (liver), embryonic (cord blood), or adult (blood or bone marrow) origin, and according to the specific activation conditions.

Retroviral vectors pseudotyped with the GaLV envelope have a broad host range, including primate, hamster and bovine cells, and may be superior to amphotropic particles to infect dog and human cells [22]. In some short-term culture studies, CFU-GM from human bone marrow or the CD34+ fraction of PBMC infected with GaLV-pseudotyped particles show significantly higher gene transfer (mean 20–25%) than those infected with amphotropic viral stocks of similar titer (mean 11–15%) [41, 46]. In one study using a vector encoding a drug resistance gene, LTC-IC yielded 28 and 35% resistant colonies, from bone marrow and PBMC, respectively, using cell-free viral stocks of GaLV-pseudotyped virions. The corresponding levels were 19 and 18% after infection with amphotropic particles [41].

Recombinant MLV virions can also be pseudotyped with VSV-G, which extends their host range and yields stable particles that withstand concentration by ultracentrifugation [23]. There are still few systematic studies on the effectiveness of VSV-G-pseudotyped virions to infect HSC. Studies limited to short-term FACS analysis [47] are tenuous because of the possible occurrence of pseudotransduction with VSV-G-pseudotyped virions [48]. An et al. [49] report transduction of human CD34+ fetal liver cells with an MLV-based retroviral vector carrying either the luciferase reporter gene or the Thy1.2 cell surface marker. By FACS analysis and quantitative PCR, they showed expression and integration of the viral DNA in thymocytes generated after transplantation of the transduced progenitor cells in SCID-hu Thy/Liv mice. Transduced thymocyte pools were shown to harbor 0.1–8.5 vector copies per cell.

Direct comparisons of gene transfer efficiency using VSV-G and amphotropic particles on cultured CD34+ cells are scarce. Akkina et al. [50] report the ability to transduce human CD34+ with an HIV-1-derived vector carrying the luciferase reporter gene under the control of the viral long terminal repeat (LTR) and pseudotyped with the VSV-G envelope glycoprotein. They demonstrate a 10-fold increase in luciferase activity after transduction of CD34+ cells with the pseudotyped virions compared to an amphotropic retrovirus stock of comparable titer. PCR analyses demonstrated maintenance of the retrovirus in CFU-GM colonies 26 days after infection. There are no published data comparing the efficacy of VSV-G to that of GaLV-pseudotyped virions in CD34+ cells or LTC-IC. In human primary T lymphocytes, GaLV-pseudotyped virions were much more effective than their VSV-G-pseudotyped counterparts at the same virus concentration [48]. However, the latter's lesser efficiency could be compensated for by using highly concentrated viral stocks.

The evaluation and optimization of human HSC transduction have often been inferred from studies in LTC-IC. While LTC-IC are generally accepted to be related to the primitive HSC, the exact nature of this relationship and thus their validity as a surrogate assay for the study of HSC as target cells for gene transfer remain unclear. In vivo readouts, which can be obtained in human-SCID mouse xenochimeras, are arguably more relevant. Recently, one group reported high-efficiency gene transfer in LTC-IC using amphotropic particles (mean of 29% in the in vitro assay), but apparently fewer ($\leq 2\%$) in the marrow of NOD/SCID mice transplanted with the transduced cells [51]. These findings call into question the actual role of LTC-IC in host repopulation as well as in their predictive value in evaluating gene transfer protocols.

There are very few studies comparing different retroviral envelopes in the context of in vivo assays. A competitive repopulation assay in baboons given CD34+ marrow cells infected with different vectors under similar conditions showed that 3 of 4 animals had consistently higher levels of provirus derived from a GaLV vector than from amphotropic vectors in both blood and marrow [52]. By semiquantitative PCR, the GaLV to amphotropic ratios were greater than 2.4. In the fourth animal, the contribution from the two vectors varied over time but the mean ratio was still about 2. These interesting results are yet to be confirmed and explored in additional studies.

Requirement for Cell Division for Murine Leukemia Virus but Not Human Immunodeficiency Virus

Retroviral vectors derived from MLV require that the target cell divide to allow proviral integration [53–55]. This is not the case for lentiviruses, like human immunodeficiency virus (HIV), which possess the ability to transport the viral preintegration complex into the nucleus in the absence of the nuclear membrane breakdown that accompanies mitosis [56]. The prospect of capitalizing on lentiviral vectors for gene transfer in resting HSC is enormous. However, the data available to date are limited to studies in macrophages, neurons, retinal cells, muscle, and liver [57–62]. These target cells are intensely metabolically active, in contrast to quiescent HSC. It is unknown at this time whether mitotically inactive HSC provide the recombinant virus with the cellular factors involved in reverse transcription, nuclear import, and retroviral integration.

Ancillary Support for Reverse Transcription

A possible factor limiting successful retroviral infection of quiescent target cells is deficient reverse transcription of the retroviral RNA. The viral enzyme reverse transcriptase requires sufficient cytoplasmic dNTP pools for efficient processive action to generate the proviral DNA. Manipulation of the intracellular dNTP concentration can have dramatic effects on virus production. For example, HIV-1 infection has been shown to be arrested at the stage of minus-strand synthesis in quiescent lymphocytes [63], potentially resulting from low concentration of cytoplasmic dNTPs [64]. Exogenous addition of dNTPs prior to infection of these resting cells has been demonstrated to improve reverse transcription of HIV-1 [64, 65]. Similar experiments with MLV virions [66] and VSV-G-pseudotyped HIV-1 virions [60] have also suggested a possible increase of transgene expression in NIH 3T3 fibroblasts and neurons, respectively.

Agrawal et al. [67] examined reverse-transcribed retroviral sequences in cytokine-stimulated CD34+CD38+ vs. CD34+CD38− hematopoietic progenitor cells 7 h after infection with a VSV-G-pseudotyped MLV-based vector. PCR analyses demonstrated preintegrated and/or integrated viral DNA in the CD34+CD38+ cell population but not in the more primitive CD34+CD38− cells. Other studies based on FACS analysis suggested that retroviral particles do not fail to bind to cytokine unstimulated or stimulated CD34+CD38− cells, nor do they fail to fuse with the target cell membrane [68]. Thus, the inability to support reverse transcription of retroviral RNA by mitotically inactive CD34+CD38− cells may represent an additional barrier to retroviral transduction.

Importance of Viral Titer

Some studies have suggested that viral titers are not rate-limiting in the transduction of HSC [69]. In view of the more recent appreciation of the importance of receptor expression, cycling status, and cellular metabolism, this notion clearly needs to be reassessed. In any event, high-titer producers present definite advantages at the level of vector production and transduction procedures.

The generation of a high-titer vector chiefly depends on the vector sequence and the selection of a good producer cell line [70]. Most cDNAs inserted in common retroviral vectors do not pose problems such as unstable viral transcripts, genomic rearrangements, or low viral titers. However, the introduction of genomic DNA, either genes or regulatory sequences, can severely reduce viral transmission. Such is the case with the introduction of the β-globin locus control region (LCR, see below) in globin vectors. Initial efforts to incorporate LCR subfragments into retroviral vectors resulted in either low titers [71], low expression [72], or unstable vectors prone to sequence rearrangements [73]. High-titer globin vectors were eventually constructed, owing to a systematic analysis of the effect of viral and genomic sequences on viral transmission [74]. Rational vector design can be achieved by detailed Southern blot analyses to quantitate vector copy number in polyclonal producer cells and generic target cells (NIH 3T3 fibroblasts) infected under standardized conditions [70, 74].

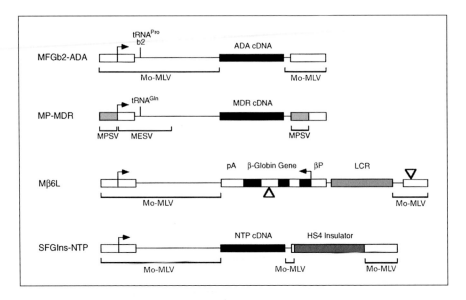

Fig. 1. Examples of vector design. The first two vectors are typical LTR-driven vectors, comprising different U3 regions and mutations in the tRNA primer binding site region [78, 79]. Mβ6L is an internal promoter vector comprising the human β-globin gene with its promoter (βP) and polyadenylation sequence (pA) and a 1-kb contraction of the human β-globin LCR [74]. The fourth vector encodes the cell surface marker NTP and the chicken HS4 globin insulator which is cloned in the 3' LTR and is thus duplicated upon integration so as to flank the vector on both ends [87].

Transgene Expression in the Progeny of Transduced Hematopoietic Stem Cells

Long Terminal Repeat-Driven Expression

In most applications of retroviral-mediated gene transfer, gene expression is under the control of constitutively active transcriptional control elements, either the retroviral LTR or internal promoters of viral or mammalian origin [see, for example, ref. 70] (fig. 1). MLV LTR have been the most extensively studied. Unique LTR have been identified in mutant MLV strains possessing different biological characteristics. For example, Moloney MLV (Mo-MLV) induces T-cell malignancies in mice, while the myeloproliferative sarcoma virus (MPSV) causes murine erythroleukemia and myeloid leukemia [75]. Murine embryonic stem cell virus (MESV/MSCV) was identified as a mutant MPSV capable of replicating in embryonal carcinoma cells [76]. While none are strictly tissue-specific, they do vary in their pattern of expression in hematopoietic cells. Thus, the MPSV LTR, which carries a few point mutations in its U3 region distinguishing it from the Mo-MLV LTR [77], expresses in all lineages, as does the Mo-MLV LTR, but provides higher expression in myeloid cells of long-term bone marrow chimeras [78]. An important feature distinguishing retroviral vectors might be the ability to express in primitive cell lines such as embryonal carcinoma (EC) and embryonal stem (ES) cells. Mutations in the MPSV and MSCV LTRs or in the tRNA primer binding site, increase LTR-driven expression in EC and/or ES cells relative to Mo-MLV [76, 79, 80]. However, it remains to be established whether improved expression in these cells correlates with improved expression in transduced HSC and in their progeny [81]. In primary T cells, these different LTRs do not appear to express differently [82].

A comprehensive analysis and comparison of different LTR-driven vectors is not a simple task as multiple mechanisms ultimately affect expression. In addition to the enhancer-promoter combination, the vectors may differ in their susceptibility to silencing mechanisms and position effects. These differences are difficult to assess in the most relevant setting of long-term bone marrow chimeras. Indeed, results obtained in small cohorts of recipients reconstituted with few transduced cells carry the risk of providing exceptions rather than illustrating the rule. One important requirement for an objective comparison is the efficient and comparable transduction of each vector in the critical subset of long-term repopulating cells. Another is the careful quantitation of protein expression and vector copy number in cells of the different hematopoietic lineages. There are very few published studies of this rigorous nature based on significant numbers of animals [78].

Two important lessons must be borne in mind when assessing gene expression for genetic therapies. The first is that expression in cultured cells can differ from that obtained in the same cells returned in vivo, as clearly demonstrated in fibroblasts [83, 84]. Therefore, demonstrating expression in cultured cells does not predict an identical outcome in the transplant recipient. The second is that transcriptional silencing is more likely to occur when transducing a primitive cell than when transducing a more differentiated cell (fig. 2). Therefore, demonstrating expression in the latter does not predict the same outcome after targeting gene transfer to a progenitor or stem cell. Transcriptional silencing [83–86] also varies according to the integration site [87], thus implicating dominant effects imparted by flanking chromatin and the 'transcriptional microenvironment'.

Locus Control Regions

The incorporation into a viral vector of an entire gene, its promoter, and enhancers, does not ensure properly regulated gene expression, as clearly demonstrated for β-globin gene retroviral vectors [88–94]. Thus, expression

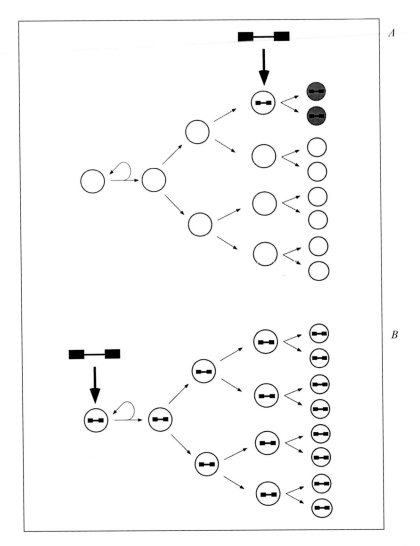

Fig. 2. Gene expression obtained with a given vector depends on the nature of the transduced cell [105]. In *A*, the globin vector integrates in an erythroid progenitor and expresses human β-globin in the erythroid progeny (filled cells). In *B*, the same vector integrates in a primitive HSC and fails to express in the same erythroid progeny (empty cells).

of the β-globin polypeptide in bone marrow chimeras is tissue-specific, but low and variable (usually varying between 0 and 2% of endogenous mouse β-major), despite the inclusion in the transcription unit of both β-globin enhancers. The advent of the concept of LCR raised the hope that expression levels of randomly integrated transgenes could be controlled in a chromosomal position-independent fashion. The LCR is thought to act on chromatin by decondensing high-order chromatin structure [95], thus creating favorable conditions for gene expression. Transcriptionally active genes are typically associated with DNAse I-sensitive regions marking active chromatin where transactivating factors are thought to easily gain access. Genes embedded in nonactive chromatin are not expressed.

The LCR in the β-globin gene cluster is associated with DNAse I hypersensitivity [96–98]. Grosveld et al. [99] and van Assendelft et al. [100] showed in cell lines and in transgenic animals that a 20-kb fragment of genomic DNA located 50–60 kb upstream of the β-globin gene confers erythroid-specific, elevated and position-independent expression upon *cis*-linked genes. Each transfected cell or transgenic line bears the construct at a different integration site. Thus, in transgenic mice bearing the 20-kb LCR, expression is detected for all integration sites in a copy-number-dependent fashion, as is the case following transfer of the entire gene cluster in yeast artificial chromosomes [101, 102].

Viral vectors cannot accommodate a 20-kb genomic sequence like the β-globin LCR. A number of studies by several investigators defined smaller segments of the LCR, or core sites, which appeared to retain part of the LCR function in transgenic mice [for a review, see ref. 103]. In the context of retroviral vectors, the HS2, HS3, and HS4 core sites of the human β-globin LCR (fig. 1) acted as powerful erythroid enhancers, but failed to provide position-independent expression [74]. In bone marrow chimeras transplanted with HSC transduced with β-globin vectors harboring the HS2, HS3, and HS4 core sites, 3 out of 12 long-term chimeras expressed some level of human β-globin in one study [104] and none out of 7 in another [105]. Additional studies are still needed to assess the potential of LCRs in the context of retroviral vectors.

The variable results obtained with different LCR fragments in different experimental systems, including transgenic mice, transfected cells, nonselected virally infected cell lines, and bone marrow chimeras, are complex, sometimes contradictory, and are discussed elsewhere [105]. These different results underscore the importance of testing transcription units designed for gene therapy in the relevant context of a viral vector in valid preclinical in vivo models.

Matrix/Scaffold Attachment Regions and Insulators

Transcription appears to be regulated by chromatin at the level of accessibility and spatial organization within the nucleus. Matrix attachment regions (MAR), also referred to as scaffold attachment regions (SAR), are DNA sequences that interact with the nuclear matrix or scaffold [106, 107]. Functionally, MAR can augment the activity of flanking enhancers and protect genes from position effects by acting as boundary elements [108].

MAR sequences located upstream and downstream to the human apolipoprotein B (ApoB) gene are boundaries between nuclease-sensitive and nuclease-resistant chromatin. In rat or human hepatoma cell lines bearing single copies of an ApoB transgene flanked by the ApoB 5′ and 3′ MAR, the level of ApoB expression is approximately 200-fold higher than in cells carrying a minimal ApoB promoter/enhancer construct [109]. However, these findings are not yet confirmed in transgenic mice [110]. When inserted in the 3′ LTR of an Mo-MLV-based internal promoter vector (fig. 1), the 1-kb ApoB 3′ MAR appeared to sustain transgene expression in skin [111], but this study lacked a quantitative comparison to a vector incorporating a control sequence cloned in the same site.

The SAR sequence associated with the human β-interferon gene (IFN-β-SAR) has also been demonstrated to enhance expression of transfected heterologous genes in vitro [112]. This element was inserted in the 3′ LTR of Mo-MLV- or MFSV-derived retroviral vectors [113]. Clonal analysis of transduced and selectively sorted CD4+ T cells showed that the SAR sequence was not able to confer position-independent expression, but could sustain expression when the T cells became quiescent. By FACS analysis, SAR-containing vectors provided a 2-fold increase in reporter gene expression in resting cells. In SCID-human xenochimeras transplanted with transduced human peripheral CD34+ cells, thymocytes harboring the SAR vector were 3–6 times brighter than thymocytes carrying the control vector. In all the latter studies, however, gene expression was not normalized to vector copy number and may have been biased by the utilization of positive selection.

In the human and chicken β-globin chromatin domains, insulator elements termed 5′ HS5 and 5′ HS4, respectively, were defined on the basis of their ability to shelter a promoter from an adjacent enhancer when they were inserted between these two elements [114–116]. In an erythropoietic cell line transduced with an Mo-MLV-based retroviral vector containing the chicken insulator HS4 (SFGins-NTP, fig. 1), we observed a doubling in the frequency of integration sites permitting expression of the vector [87]. Additional studies will be necessary to establish whether insulators actually reduce variability of expression after random chromosomal integration of retroviral vectors and whether they can offset transcriptional silencing.

Conclusion

Despite constant progress in the purification of HSC and in the development of new assays for studying virally transduced HSC, efficient retroviral-mediated gene transfer in human HSC remains an elusive goal. A number of questions remain to be answered in order to build performant vectors for efficient transduction and regulated transgene expression in the hematopoietic tissue. It is not yet clear which retroviral receptors are expressed by HSC, how receptor expression can be optimized, and what retroviral envelopes are best suited for HSC transduction. Nor is it known at this time whether the requirement for cell division to achieve MLV integration will be surmounted by lentiviral vectors. The regulation and durability of transgene expression in vivo in HSC and/or their progeny also pose major challenges. The mechanisms leading to position effects and transcriptional silencing appear to be multiple, and it remains to be established to what extent modifications in viral elements and the incorporation of chromatin regulators will permit regulated and sustained transgene expression in vivo. Likewise, a better understanding of HSC biology is necessary to improve our ability to identify, expand and transduce these cells without impairing their developmental capacity.

However, the transduction of less primitive primary progenitor cells has greatly benefited from efforts to genetically modify HSC. Thus, the technology currently available may be adequate to meet goals that require the transduction of progenitor cells, such as the transient supply of drug-resistant myeloid cells.

Acknowledgment

Our work is supported by NIH grants CA-59350 and HL-57612 (MS).

References

1. Brecher G, Bookstein N, Redfearn W, Necas E, Pallavicini MG, Cronkite EP: Self-renewal of the long-term repopulating stem cell. Proc Natl Acad Sci USA 1993;90:6028–6031.
2. Okada S, Nagayoshi K, Nakauchi H, Nishikawa S, Miura Y, Suda T: Sequential analysis of hematopoietic reconstitution achieved by transplantation of hematopoietic stem cells. Blood 1993; 81:1720–1725.
3. Fleming WH, Alpern EJ, Uchida N, Ikuta K, Sprangrude GJ, Weissman IL: Functional heterogeneity is associated with the cell cycle status of murine hematopoietic stem cells. J Cell Biol 1993;122: 897–902.
4. Peault B, Weissman IL, Buckle AM, Tsukamoto A, Baum C: Thy-1-expressing CD34+ human cells express multiple hematopoietic potentialities in vitro and in SCID-hu mice. Nouv Rev Fr Hématol 1993;35:91–93.
5. Srour EF, Zanjani ED, Cornetta K, Traycoff CM, Flake AW, Hedrick M, Brandt JE, Leemhuis T, Hoffman R: Persistence of human multilineage, self-renewing lymphohematopoietic stem cells in chimeric sheep. Blood 1993;82:3333–3342.
6. To LB, Haylock DN, Simmons PJ, Juttner CA: The biology and clinical uses of blood stem cells. Blood 1997;89:2233–2258.
7. Issaad C, Croisille L, Katz A, Vainchenker W, Coulombel L: A murine stromal cell line allows the proliferation of very primitive human CD34++/CD38− progenitor cells in long-term cultures and semisolid assays. Blood 1993;81:2916–2924.

8   Uchida N, Combs J, Chen S, Zanjani E, Hoffman R, Tsukamoto A: Primitive human hematopoietic cells displaying differential efflux of the rhodamine 123 dye have distinct biological activities. Blood 1996;88:1297–1305.
9   Goodell MA, Brose K, Paradis G, Conner AS, Mulligan RC: Isolation and functional properties of murine hematopoietic stem cells that are replicating in vivo. J Exp Med 1996;183:1797–1806.
10   Goodell MA, Rosenzweig M, Kim H, Marks DF, DeMaria M, Paradis G, Grupp SA, Sieff CA, Mulligan RC, Johnson RP: Dye efflux studies suggest that hematopoietic stem cells expressing low or undetectable levels of CD34 antigen exist in multiple species. Nat Med 1997;3:1337–1345.
11   Neben S, Anklesaria P, Greenberger J, Mauch P: Quantitation of murine hematopoietic stem cells in vitro by limiting dilution analysis of cobblestone area formation on a clonal stromal cell line [see comments]. Exp Hematol 1993;21:438–443.
12   Tucker D, Bol S, Kannourakis G: Characterization of stroma-adherent colony-forming cells: A clonogenic assay for early hemopoietic cells? Exp Hematol 1993;21:469–474.
13   Goldstein NI, Moore MAS, Allen C, Tackney C: A human spleen cell line, immortalized with SV40 T-antigen, will suppport the growth of CD34+ long-term culture-initiating cells. Mol Cell Different 1993;1:301–321.
14   Hao QL, Thiemann FT, Petersen D, Smogorzewska EM, Crooks GM: Extended long-term culture reveals a highly quiescent and primitive human hematopoietic progenitor population. Blood 1996; 88:3306–3313.
15   Fraswer CC, Szilvassy SJ, Eaves CJ, Humphries RK: Proliferation of totipotent hematopoietic stem cells in vitro with retention of long-term competitive in vivo reconstituting ability. Proc Natl Acad Sci USA 1992;89:1968–1972.
16   Miller AD: Cell-surface receptors for retroviruses and implications for gene transfer. Proc Natl Acad Sci USA 1996;93:11407–11413.
17   Miller AD, Buttimore C: Redesign of retrovirus packaging cell lines to avoid recombination leading to helper virus production. Mol Cell Biol 1986;6:2895–2902.
18   Danos O, Mulligan RC: Safe and efficient generation of recombinant retroviruses with amphotropic and ecotropic host ranges. Proc Natl Acad Sci USA 1988;85:6460–6464.
19   Markowitz DG, Goff SP, Bank A: Safe and efficient ecotropic and amphotropic packaging lines for use in gene transfer experiments. Trans Assoc Am Physicians 1988;101:212–218.
20   Forestell SP, Dando JS, Chen J, de Vries P, Bohnlein E, Rigg RJ: Novel retroviral packaging cell lines: Complementary tropisms and improved vector production for efficient gene transfer. Gene Ther 1997;4:600–610.
21   Kinsella TM, Nolan GP: Episomal vectors rapidly and stably produce high-titer recombinant retrovirus. Hum Gene Ther 1996;7:1405–1413.
22   Miller AD, Garcia JV, von Suhr N, Lynch CM, Wilson C, Eiden MV: Construction and properties of retrovirus packaging cells based on gibbon ape leukemia virus. J Virol 1991;65:2220–2224.
23   Burns JC, Friedmann T, Driever W, Burrascano M, Yee JK: Vesicular stomatitis virus G glycoprotein pseudotyped retroviral vectors: Concentration to very high titer and efficient gene transfer into mammalian and nonmammalian cells [see comments]. Proc Natl Acad Sci USA 1993;90:8033–8037.
24   Miller DG, Edwards RH, Miller AD: Cloning of the cellular receptor for amphotropic murine retroviruses reveals homology to that for gibbon ape leukemia virus. Proc Natl Acad Sci USA 1994; 91:78–82.
25   van Zeijl M, Johann SV, Closs E, Cunningham J, Eddy R, Shows TB, O'Hara B: A human amphotropic retrovirus receptor is a second member of the gibbon ape leukemia virus receptor family. Proc Natl Acad Sci USA 1994;91:1168–1172.
26   Kavanaugh MP, Miller DG, Zhang W, Law W, Kozak SL, Kabat D, Miller AD: Cell-surface receptors for gibbon ape leukemia virus and amphotropic murine retrovirus are inducible sodium-dependent phosphate symporters. Proc Natl Acad Sci USA 1994;91:7071–7075.
27   Schlegel R, Tralka TS, Willingham MC, Pastan I: Inhibition of VSV binding and infectivity by phosphatidylserine: Is phosphatidylserine a VSV-binding site? Cell 1983;32:639–664.
28   Miller AD, Chen F: Retroovirus packaging cells based on 10A1 murine leukemia virus for production of vectors that use multiple receptors for cell entry. J Virol 1996;70:5564–5571.

29 Ory DS, Neugeboren BA, Mulligan RC: A stable human-derived packaging cell line for production of high titer retrovirus/vesicular stomatitis virus G pseudotypes. Proc Natl Acad Sci USA 1996;93:11400–11406.
30 Orlic D, Girard LJ, Jordan CT, Anderson SM, Cline AP, Bodine DM: The level of mRNA encoding the amphotropic retrovirus receptor in mouse and human hematopoietic stem cells is low and correlates with the efficiency of retrovirus transduction. Proc Natl Acad Sci USA 1996;93:11097–11102.
31 Bienzle D, Abrams-Ogg AC, Kruth SA, Ackland-Snow J, Carter RF, Dick JE, Jacobs RM, Kamel-Reid S, Dube ID: Gene transfer into hematopoietic stem cells: Long-term maintenance of in vitro activated progenitors without marrow ablation. Proc Natl Acad Sci USA 1994;91:350–354.
32 Kiem HP, Darovsky B, von Kalle C, Goehle S, Stewart D, Graham T, Hackman R, Appelbaum FR, Deeg HJ, Miller AD, et al: Retrovirus-mediated gene transduction into canine peripheral blood repopulating cells. Blood 1994;83:1467–1473.
33 Kiem HP, Darovsky B, von Kalle C, Goehle S, Graham T, Miller AD, Storb R, Schuening FG: Long-term persistence of canine hematopoietic cells genetically marked by retrovirus vectors. Hum Gene Ther 1996;7:89–96.
34 Kantoff PW, Gillio AP, McLachlin JR, Bordignon C, Eglitis MA, Kernan NA, Moen RC, Kohn DB, Yu SF, Karson E, et al: Expression of human adenosine deaminase in nonhuman primates after retrovirus-mediated gene transfer. J Exp Med 1987;166:219–234.
35 Bodine DM, Moritz T, Donahue RE, Luskey BD, Kessler SW, Martin DI, Orkin SH, Nienhuis AW, Williams DA: Long-term in vivo expression of a murine adenosine deaminase gene in rhesus monkey hematopoietic cells of multiple lineages after retroviral mediated gene transfer into CD34+ bone marrow cells. Blood 1993;82:1975–1980.
36 Brenner MK, Rill DR, Holladay MS, Heslop HE, Moen RC, Buschle M, Krance RA, Santana VM, Andersen WF, Ihle JN: Gene marking to determine whether autologous marrow infusion restores long-term haemopoiesis in cancer patients. Lancet 1993;342:1134–1137.
37 Kohn DB, Weinberg KI, Nolta JA, Heiss LN, Lenarsky C, Crooks GM, Hanley ME, Annett G, Brooks JS, el-Khoureiy A, et al: Engraftment of gene-modified umbilical cord blood cells in neonates with adenosine deaminase deficiency. Nat Med 1995;1:1017–1023.
38 Bordignon C, Notarangelo LD, Nobili N, Ferrari G, Casorati G, Panina P, Mazzolari E, Maggioni D, Rossi C, Servida P, et al: Gene therapy in peripheral blood lymphocytes and bone marrow for ADA-immunodeficient patients. Science 1995;270:470–475.
39 Dunbar CE, Cottler-Fox M, O'Shaughnessy JA, Doren S, Carter C, Berenson R, Brown S, Moen RC, Greenblatt J, Stewart FM, et al: Retrovirally marked CD34-enriched peripheral blood and bone marrow cells contribute to long-term engraftment after autologous transplantation. Blood 1995;85:3048–3057.
40 Scott-Taylor TH, Gallardo HF, Gansbacher B, Sadelain M: Adenovirus facilitated infection of human cells with ecotropic retrovirus. Gene Ther 1998;5:621–629.
41 von Kalle C, Kiem HP, Goehle S, et al: Increased gene transfer into human hematopoietic progenitor cells by extended in vitro exposure to a pseudotyped retroviral vector. Blood 1994;84:2890–2897.
42 Crooks GM, Kohn DB: Growth factors increase amphotropic retrovirus binding to human CD34+ bone marrow progenitor cells. Blood 1993;82:3290–3297.
43 Richardson C, Ward M, Podda S, Bank A: Mouse fetal liver cells lack functional amphotropic retroviral receptors. Blood 1994;84:433–439.
44 Bauer TR Jr, Miller AD, Hickstein DD: Improved transfer of the leukocyte integrin CD18 subunit into hematopoietic cell lines by using retroviral vectors having a gibbon ape leukemia virus envelope. Blood 1995;86:2379–2387.
45 Orlic D, Girard LJ, Anderson SM, Do BK, Seidel NE, Jordan CT, Bodine DM: Transduction efficiency of cell lines and hematopoietic stem cells correlates with retrovirus receptor mRNA levels. Stem Cells 1997;15(suppl 1):23–28; discussion 28–29.
46 Dybing J, Lynch CM, Hara P, Jurus L, Kiem HP, Anklesaria P: GaLV pseudotyped vectors and cationic lipids transduce human CD34+ cells. Hum Gene Ther 1997;8:1685–1694.
47 Reiser J, Harmison G, Kluepfel-Stahl S, Brady RO, Karlsson S, Schubert M: Transduction of nondividing cells using pseudotyped defective high-titer HIV type 1 particles. Proc Natl Acad Sci USA 1996;93:15266–15271.

48 Gallardo HF, Tan C, Ory D, Sadelain M: Recombinant retroviruses pseudotyped with the vesicular stomatitis virus G glycoprotein mediate both stable gene transfer and pseudotransduction in human peripheral blood lymphocytes. Blood 1997;90:952–957.

49 An DS, Koyanagi Y, Zhao JQ, Akkina R, Bristol G, Yamamoto N, Zack JA, Chen IS: High-efficiency transduction of human lymphoid progenitor cells and expression in differentiated T cells. J Virol 1997;71:1397–1404.

50 Akkina RK, Walton RM, Chen ML, Li QX, Planelles V, Chen IS: High-efficiency gene transfer into CD34+ cells with a human immunodeficiency virus type 1-based retroviral vector pseudotyped with vesicular stomatitis virus envelope glycoprotein G. J Virol 1996;70:2581–2585.

51 Larochelle A, Vormoor J, Hanenberg H, Wang JC, Bhatia M, Lapidot T, Moritz T, Murdoch B, Xiao XL, Kato I, Williams DA, Dick JE: Identification of primitive human hematopoietic cells capable of repopulating NOD/SCID mouse bone marrow: Implications for gene therapy. Nat Med 1996;2:1329–1337.

52 Kiem HP, Heyward S, Winkler A, Potter J, Allen JM, Miller AD, Andrews RG: Gene transfer into marrow repopulating cells: Comparison between amphotropic and gibbon ape leukemia virus pseudotyped retroviral vectors in a competitive repopulation assay in baboons. Blood 1997;90:4638–4645.

53 Miller DG, Adam MA, Miller AD: Gene transfer by retrovirus vectors occurs only in cells that are actively replicating at the time of infection [published erratum appears in Mol Cell Biol 1992; 12:433]. Mol Cell Biol 1990;10:4239–4242.

54 Roe T, Reynolds TC, Yu G, Brown PO: Integration of murine leukemia virus DNA depends on mitosis. Embo J 1993;12:2099–2108.

55 Hajihosseini M, Iavachev L, Price J: Evidence that retroviruses integrate into post-replication host DNA. Embo J 1993;12:4969–4974.

56 Lewis PF, Emerman M: Passage through mitosis is required for oncoretroviruses but not for the human immunodeficiency virus. J Virol 1994;68:510–516.

57 Zufferey R, Nagy D, Mandel RJ, Naldini L, Trono D: Multiple attenuated lentiviral vector achieves efficient gene delivery in vivo. Nat Biotechnol 1997;15:871–875.

58 Naldini L, Blomer U, Gage FH, Trono D, Verma IM: efficient transfer, integration, and sustained long-term expression of the transgene in adult rat brains injected with a lentiviral vector. Proc Natl Acad Sci USA 1996;93:11382–11388.

59 Naldini L, Blomer U, Gallay P, Ory D, Mulligan R, Gage FH, Verma IM, Trono D: In vivo gene delivery and stable transduction of nondividing cells by a lentiviral vector [see comments]. Science 1996;272:263–267.

60 Blomer U, Naldini L, Kafri T, Trono D, Verma IM, Gage FH: Highly efficient and sustained gene transfer in adult neurons with a lentivirus vector. J Virol 1997;71:6641–6649.

61 Miyoshi H, Takahashi M, Gage FH, Verma IM: Stable and efficient gene transfer into the retina using an HIV-based lentiviral vector. Proc Natl Acad Sci USA 1997;94:10319–10323.

62 Kafri T, Blomer U, Peterson DA, Gage FH, Verma IM: Sustained expression of genes delivered directly into liver and muscle by lentiviral vectors. Nat Genet 1997;17:314–317.

63 Zack JA, Haislip AM, Krogstad P, Chen IS: Incompletely reverse-transcribed human immunodeficiency virus type 1 genomes in quiescent cells can function as intermediates in the retroviral life cycle. J Virol 1992;66:1717–1725.

64 Gao WY, Cara A, Gallo RC, Lori F: Low levels of deoxynucleotides in peripheral blood lymphocytes: A strategy to inhibit human immunodeficiency virus type 1 replication. Proc Natl Acad Sci USA 1993;90:8925–8928.

65 Meyerhans A, Vartanian JP, Hultgren C, Plikat U, Karlsson A, Wang L, Eriksson S, Wain-Hobson S: Restriction and enhancement of human immunodeficiency virus type 1 replication by modulation of intracellular deoxynucleoside triphosphate pools. J Virol 1994;68:535–540.

66 Zhang H, Duan LX, Dornadula G, Pomerantz RJ: Increasing transduction efficiency of recombinant murine retrovirus vectors by initiation of endogenous reverse transcription: Potential utility for genetic therapies. J Virol 1995;69:3929–3932.

67 Agrawal YP, Agrawal RS, Sinclair AM, Young D, Maruyama M, Levine F, Ho AD: Cell-cycle kinetics and VSV-G pseudotyped retrovirus-mediated gene transfer in blood-derived CD34+ cells. Exp Hematol 1996;24:738–474.

68 Sinclair AM, Agrawal YP, Elbar E, Agrawal R, Ho AD, Levine F: Interaction of vesicular stomatitis virus-G pseudotyped retrovirus with CD34+ and CD34+ CD38– hematopoietic progenitor cells. Gene Ther 1997;4:918–927.

69 Van Beusechem VW, Bakx TA, Kaptein LC, Bart-Baumeister JA, Kukler A, Braakman E, Valerio D: Retrovirus-mediated gene transfer into rhesus monkey hematopoietic stem cells: The effect of viral titers on transduction efficiency. Hum Gene Ther 1993;4:239–247.

70 Riviere I, Sadelain M: Methods for the construction of retroviral vectors and the generation of high-titer producers; in Robbins PD (ed): Methods in Molecular Medicine, Gene Therapy Protocols. Totowa, Humana Press, 1997, pp 59–78.

71 Plavec I, Papayannopoulou T, Maury C, Meyer F: A human beta-globin gene fused to the human beta-globin locus control region is expressed at high levels in erythroid cells of mice engrafted with retrovirus-transduced hematopoietic stem cells. Blood 1993;81:1384–1392.

72 Chang JC, Liu D, Kan YW: A 36-base-pair core sequence of locus control region enhances retrovirally transferred human beta-globin gene expression. Proc Natl Acad Sci USA 1992;89:3107–3110.

73 Novak U, Harris EA, Forrester W, Groudine M, Gelinas R: High-level beta-globin expression after retroviral transfer of locus activation region-containing human beta-globin gene derivatives into murine erythroleukemia cells. Proc Natl Acad Sci USA 1990;87:3386–3390.

74 Sadelain M, Wang CH, Antoniou M, Grosveld F, Mulligan RC: Generation of a high-titer retroviral vector capable of expressing high levels of the human beta-globin gene. Proc Natl Acad Sci USA 1995;92:6728–6732.

75 Ostertag W, Vehmeyer K, Fagg B, Pragnell IB, Paetz W, Le Bousse MC, Smadja-Joffe F, Klein B, Jasmin C, Eisen H: Myeloproliferative virus, a cloned murine sarcoma virus with spleen focus-forming properties in adult mice. J Virol 1980;33:573–582.

76 Grez M, Akgun E, Hilberg F, Ostertag W: Embryonic stem cell virus, a recombinant murine retrovirus with expression in embryonic stem cells. Proc Natl Acad Sci USA 1990;87:9202–9206.

77 Stocking C, Kollek R, Bergholz U, Ostertag W: Long terminal repeat sequences impart hematopoietic transformation properties to the myeloproliferative sarcoma virus. Proc Natl Acad Sci 1985;82:5746–5750.

78 Riviere I, Brose K, Mulligan RC: Effects of retroviral vector design on expression of human adenosine deaminase in murine bone marrow transplant recipients engrafted with genetically modified cells. Proc Natl Acad Sci USA 1995;92:6733–6737.

79 Baum C, Hegewisch-Becker S, Eckert HG, Stocking C, Ostertag W: Novel retroviral vectors for efficient expression of the multidrug resistance (mdr-1) gene in early hematopoietic cells. J Virol 1995;69:7541–7547.

80 Baum C, Eckert HG, Stockschlader M, Just U, Hegewisch-Becker S, Hildinger M, Uhde A, John J, Ostertag W: Improved retroviral vectors for hematopoietic stem cell protection and in vivo selection. J Hematother 1996;5:323–329.

81 Hawley RG, Fong AZ, Burns BF, Hawley TS: Transplantable myeloproliferative disease induced in mice by an interleukin 6 retrovirus. J Exp Med 1992;176:1149–1163.

82 Plavec I, Agarwal M, Ho KE, Pineda M, Auten J, Baker J, Matsuzaki H, Escaich S, Bonyhadi M, Bohnlein E: High transdominant RevM10 protein levels are required to inhibit HIV-1 replication in cell lines and primary TR cells: Implication for gene therapy of AIDS. Gene Ther 1997;4:128–139.

83 Palmer TD, Rosman GJ, Osborne WR, Miller AD: Genetically modified skin fibroblasts persist long after transplantation but gradually inactivate introduced genes. Proc Natl Acad Sci USA 1991;88:1330–1334.

84 Luo XN, Reddy JC, Yeyati PL, Idris AH, Hosono S, Haber DA, Licht JD, Atweh GF: The tumor suppressor gene WT1 inhibits ras-mediated transformation. Oncogene 1995;11:743–750.

85 Challita PM, Kohn DB: Lack of expression from a retroviral vector after transduction of murine hematopoietic stem cells is associated with methylation in vivo. Proc Natl Acad Sci USA 1994;91:2567–2571.

86 Krall WJ, Skelton DC, Yu XJ, Riviere I, Lehn P, Mulligan RC, Kohn DB: Increased levels of spliced RNA account for augmented expression from the MFG retroviral vector in hematopoietic cells. Gene Ther 1996;3:37–48.

87 Rivella S, Callegari J, Sadelain M: Characterization of a retroviral vector bearing chicken HS4 element: Effect on position effect variegation. Seattle, Am Soc Gene Ther, 1998.
88 Cone RD, Weber-Benarous A, Baorto D, Mulligan RC: Regulated expression of a complete human beta-globin gene encoded by a transmissible retrovirus vector. Mol Cell Biol 1987;7:887–897.
89 Karlsson S, Papayannopoulou T, Schweiger SG, Stamatoyannopoulos G, Nienhuis AW: Retroviral-mediated transfer of genomic globin genes leads to regulated production of RNA and protein. Proc Natl Acad Sci USA 1987;84:2411–2415.
90 Miller AD, Bender MA, Harris EA, Kaleko M, Gelinas RE: Design of retrovirus vectors for transfer and expression of the human beta-globin gene [published erratum appears in J Virol 1989;63:1493]. J Virol 1988;62:4337–4345.
91 Bender MA, Miller AD, Gelinas RE: Expression of the human beta-globin gene after retroviral transfer into murine erythroleukemia cells and human BFU-E cells. Mol Cell Biol 1988;8:1725–1735.
92 Karlsson S, Bodine DM, Perry L, Papayannopoulou T, Nienhuis AW: Expression of the human beta-globin gene following retroviral-mediated transfer into multipotential hematopoietic progenitors of mice. Proc Natl Acad Sci USA 1988;85:6062–6066.
93 Bodine DM, Karlsson S, Nienhuis AW: Combination of interleukins 3 and 6 preserves stem cell function in culture and enhances retrovirus-mediated gene transfer into hematopoietic stem cells. Proc Natl Acad Sci USA 1989;86:8897–8901.
94 Bender MA, Gelinas RE, Miller AD: A majority of mice show long-term expression of a human beta-globin gene after retrovirus transfer into hematopoietic stem cells. Mol Cell Biol 1989;9:1426–1434.
95 Felsenfeld G: Chromatin as an essential part of the transcritpional mechanism. Nature 1992;355:219–224.
96 Tuan D, Solomon W, Li Q, London IM: The 'beta-like-globin' gene domain in human erythroid cells. Proc Natl Acad Sci USA 1985;82:6384–6388.
97 Forrester WC, Takegawa S, Papayannopoulou T, Stamatoyannopoulos G, Groudine M: Evidence for a locus activation region: The formation of developmentally stable hypersensitive sites in globin-expressing hybrids. Nucleic Acids Res 1987;15:10159–10177.
98 Forrester WC, Epner E, Driscoll MC, Enver T, Brice M, Papayannopoulou T, Groudine M: A deletion of the human beta-globin locus activation region causes a major alteration in chromatin structure and replication across the entire beta-globin locus. Genes Dev 1990;4:1637–1649.
99 Grossveld F, van Assendelft GB, Greaves DR, Kollias G: Position-independent, high-level expression of the human beta-globin gene in transgenic mice. Cell 1987;51:975–985.
100 van Assendelft GB, Hanscombe O, Grosveld F, Greaves DR: The beta-globin dominant control region activates homologous and heterologous promoters in a tissue-specific manner. Cell 1989;56:969–677.
101 Peterson KR, Clegg CH, Huxley C, Josephson BM, Haugen HS, Furukawa T, Stamatoyannopoulos G: Transgenic mice containing a 248-kb yeast artificial chromosome carrying the human beta-globin locus display proper developmental control of human globin genes. Proc Natl Acad Sci USA 1993;90:7593–7597.
102 Gaensler KM, Kitamura M, Kan YW: Germ-line transmission and developmental regulation of a 150-kb yeast artificial chromosome containing the human beta-globin locus in transgenic mice. Proc Natl Acad Sci USA 1993;90:11381–11385.
103 Sadelain M: Genetic treatment of the haemoglobinopathies: Recombinations and new combinations. Br J Haematol 1997;98:247–253.
104 Raftopoulos H, Ward M, Leboulch P, Bank A: long-term transfer and expression of the human beta-globin gene in a mouse transplant model. Blood 1997;90:3414–3422.
105 Rivella S, Sadelain M: Genetic treatment of severe hemoglobinopathies: The combat against trans-gene variegation and transgene silencing. Semin Hematol 1998;35:112–125.
106 Mirkovitch J, Mirault ME, Laemmli UK: Organization of the higher-order chromatin loop: Specific DNA attachment sites on nuclear scaffold. Cell 1984;39:223–232.
107 Cockerill PN, Garrard WT: Chromosomal loop anchorage of the kappa immunoglobulin gene occurs next to the enhancer in a region containing topoisomerase II sites. Cell 1986;44:273–282.

108 Gerasimova TI, Corces VG: Boundary and insulator elements in chromosomes. Curr Opin Genet Dev 1996;6:185–192.
109 Kalos M, Fournier RE: Position-independent transgene expression mediated by boundary elements from the apolipoprotein B chromatin domain. Mol Cell Biol 1995;15:198–207.
110 Wang DM, Taylor S, Levy-Wilson B: Evaluation of the function of the human apolipoprotein B gene nuclear matrix association regions in transgenic mice. J Lipid Res 1996;37:2117–2124.
111 Deng H, Lin Q, Khavari PA: Sustainable cutaneous gene delivery. Nat Biotechnol 1997;15:1388–1391.
112 Klehr D, Maass K, Bode J: Scaffold-attached regions from the human interferon beta domain can be used to enhance the stable expression of genes under the control of various promoters. Biochemistry 1991;30:1264–1270.
113 Agarwal MATW, Morel F, Chen J, Bohnlein E, Plavec I: Scaffold attachment region-mediated enhancement of retroviral vector expression in primary T cells. J Virol 1998;72:3720–3728.
114 Chung JH, Whiteley M, Felsenfeld G: A 5′ element of the chicken beta-globin domain serves as an insulator in human erythroid cells and protects against position effect in Drosophila. Cell 1993; 74:505–514.
115 Li Q, Stamatoyannopoulos G: Hypersensitive site 5 of the human beta locus control region functions as a chromatin insulator. Blood 1994;84:1399–1401.
116 Chung JH, Bell AC, Felsenfeld G: Characterization of the chicken beta-globin insulator. Proc Natl Acad Sci USA 1997;94:575–580.

M. Sadelain, MD, PhD, Department of Human Genetics,
MSKCC, Box 182, 1275 York Avenue, New York, NY 10021 (USA)
Fax +1 (212) 717 3374, E-Mail m-sadelain@ski.mskcc.org

Optimizing Conditions for Gene Transfer into Human Hematopoietic Cells

Malcolm A.S. Moore, Karen L. MacKenzie

Sloan-Kettering Institute for Cancer Research, New York, N.Y., USA

Introduction

Clinical trials of gene therapy were first begun nearly a decade ago. Despite significant progress in understanding the limitations of gene transfer, clinical protocols suffer from poor efficiency of gene delivery and transient transgene expression. Gene transfer to hematopoietic stem cells (HSC) is an attractive therapy for a number of diseases, both because of the relative ease of obtaining and transplanting such cells, and because of the self-renewal and multilineage differentiation potential of HSC. Genetic diseases for which stem cell gene therapy has been used or proposed include severe immunodeficiency disease associated with adenosine deaminase deficiency [1–3], the hemoglobinopathies with structural defects in α- or β-globin genes (thalassemias, sickle cell anemia) [4–7], Gaucher's disease with defects in glucocerebrosidase [8–10], leukocyte adhesion deficiency with defects in the leukocyte integrin CD18 subunit [11], and chronic granulomatous disease with defects in gp91-phox expression [12]. In acquired disease, gene therapy has been proposed to treat acquired immune deficiency syndrome (AIDS) [13]. Vectors encoding anti-HIV ribozymes [14–17], overexpression of 'decoy' genes [13, 18], or expression of a transdominant mutant of HIV-rev which inhibits HIV replication [19, 20], are all under evaluation. Transduction of HSC with genes that confer resistance to myelosuppressive chemotherapy has a potential role in cancer therapy. The use of the multidrug resistance (MDR) gene which confers resistance to anthracyclines, *Vinca* alkaloids, podophyllins, and taxol is discussed in detail elsewhere in this volume, as is the use of mutant dihydrofolate reductase (mDHFR) genes to confer methotrexate or trimetrexate resistance. Methylating agents (procarbazine) and chlorethylating agents (nitrosoureas) have myelosuppres-

Table 1. Variables in retroviral transduction and evaluation

Virus	Promoter	Envelope	Marker	In vitro assay	In vivo assay
MoMLV	MoMLV	Ampho	Neo R-G418 res.	CFU-GM/BFU-E	SCID-Hu
Lentivirus	VSV-G	Galv	MDR-taxol res.	HPP-CFU	NOD/SCID
	MPSV	MSCV	NGFR	mDHFR-Mtx res.	Bnx
			β-Gal.	LTC-IC-2^0 CFU	fetal sheep
			GFP	CD34+ FACS	
			mHSA	CAFC	

For abbreviations, see text.

sive, dose-limiting toxicities, and protection can be conferred by expression of O^6-alkylguanine-DNA-alkyltransferase [21]. Cyclophosphamide resistance can be conferred by gene transfer of aldehyde dehydrogenase [22].

The vast majority of clinical and basic research studies have used vectors based upon murine leukemia viruses (MLVs). This has required development of strategies to optimize proliferation of target cells as a necessary prerequisite for retroviral transduction. The recent development of vectors based upon lentiviruses, particularly HIV-based vectors, represents a potential advance, in that transduction of nonproliferating cells is now possible [23–26]. However, it remains likely that retroviral transduction requires cells to be metabolically active, and deeply quiescent cells may still prove refractory to transduction with lentivirus. Furthermore, issues concerning viral envelopes, viral receptor expression, and stability of transgene expression remain valid concerns in both MLV and lentiviral transductions. In any attempt to draw conclusions concerning optimal strategies for gene transfer into hematopoietic cells, consideration must be given to the multiple variables in methodology, involving vector design, viral envelope, transduction protocol, marker system used, cell target (tables 1, 2). In this chapter, each of these variables is discussed, and an attempt is made to conclude which are the most critical in determining optimal stable transduction of HSC with in vivo, long-term repopulating potential.

Source of Hematopoietic Stem Cells for Transduction

Tissues rich in HSC include fetal liver (FL), umbilical cord blood (CB), adult bone marrow (BM) and apheresed peripheral blood (mPB) obtained following a course of chemotherapy, generally with cyclophosphamide, or

Table 2. Additional variables

Cell source	BM, mobilized PB, CB, FL, CD34+, MNCs
Cytokine priming	none or 1–5 days prior to virus addition
Cytokines	IL-3, IL-6, KL, Flk-2L, Tpo, IL-11, GM-CSF, G-CSF
Cytokine dose	varies by order of magnitude in range 10–100 ng/ml
Serum	fetal calf serum (10–30%), anti-TGFβ, autologous serum/plasma, serum-free media (ex vivo)
Polycations	polybrene, protamine (4–8 ug/ml), none
Virus stock	fresh/frozen, viral titer ($1 \times 10^5 - 1 \times 10^7$/ml), m.o.i. (1–10)
Transduction method	Co-culture, supernatant with variable cycles of exposures to virus, static, centrifugation, perfusion, fibronectin/Retronectin or stromal co-culture

following a 5- to 7-day course of G-CSF or GM-CSF, or a combination of CSF and cyclophosphamide. Cells from each source have both advantages and disadvantages as gene transfer targets. From an ontogenetic perspective, stem cells in FL and CB have greater self-renewal potential, are easier to maintain in culture, and may be activated into cycle more readily than adult BM or mPB cells. Also, CB and FL-derived long-term culture-initiating cells (LTC-IC) and SCID repopulating cells (SRC) expand more readily with any given cytokine combination than do adult cells [27, 28]. The engraftment efficiency of CD34+ cells from CB exceeds that of adult CD34+ cells by an order of magnitude in NOD/SCID mice [29], and the number of CB CD34+ cells required to give comparable platelet and neutrophil recovery in human transplantation is tenfold less than the requirement for adult mPB cells (2×10^5/kg versus 2×10^6/kg). Direct comparison of transduction efficiency of in vitro, hematopoietic colony-forming cells (CFC) from different sources showed a modest differential in susceptibility to transduction: CB>mPB>BM [30]. Others have shown a higher efficiency of gene transfer to CFC from CB and mPB from infants than from adult BM populations [31].

An additional advantage of CD34+ cells from fetal and neonatal sources is that gene expression appears to be more stable in progeny of stem cells from FL and CB than in cells derived from adult tissues. We, and others, have observed silencing of transgenes driven by the Moloney murine leukemia virus (MoMLV) long terminal repeat (LTR) in adult mPB stem cells [32, 33]. In our own investigations, CD34+ cells from FL, CB and mPB, were transduced with a MoMLV-based retroviral vector carrying a mutated nerve growth factor receptor (NGFR). NGFR+ cells were sorted by flow cytometry and cocultured on stromal cells. Weekly fluorescence-activated cell scanner (FACS) analysis of NGFR+ sorted cells revealed that NGFR expressing cells (as a percentage of the total population) fell progressively. The decrease in expression was less dramatic in cultures from adult cells than from cultures initiated with CB or FL cells [32]. In cultures started with mPB, NGFR expression fell from >95% at the outset to <5% after 6 weeks, whereas expression in CB cell cultures fell to 33%, and expression in FL cultures was 60% at week 7.

A potential disadvantage of these fetal and neonatal sources is that the restricted number of CD34+ cells obtained limits their use for gene therapy of adults. However, this limitation may be overcome with improved ex vivo expansion strategies that retain long-term repopulating capacity. An advantage of FL and BM cells, particulary BM from children, is the higher cell cycle activity within the CD34+ population, including stem cells, in contrast to the predominantly noncycling populations in CB and mPB. It is significant that the greatest success to date in obtaining long-term expression of a retroviral gene in HSC in a clinical trial utilized BM from young individuals post chemotherapy [34]. However, the potential of using regenerating adult BM following chemotherapy and in vivo cytokine stimulation has not been fully exploited.

Whole mononuclear cells (MNC) from BM, PB, and CB can also be used as an HSC source for gene transfer following cytokine priming. However, the presence of CD34– cells appears to inhibit HSC activation. Hence, cytokine priming of MNC is less efficient than direct stimulation of enriched progenitor populations. Furthermore, in order to obtain a multiplicity of infection (m.o.i.)>1 relative to CD34+ cells, large amounts of virus are required. CD34+ cells can be enriched by either immunomagnetic depletion with a cocktail of antibodies to CD epitopes expressed on differentiated cells, or by positive selection using immunomagnetic beads or FACS sorting. HSC may be further enriched by depletion of CD38+ cells or selection for Thy-1 or HLA-DRlo cells. The existence of a CD34– stem cell, arguably more primitive and certainly more quiescent than the CD34+ stem cells, has been demonstrated using transplantation to NOD/SCID mice [34] and to fetal lambs [35]. An evaluation of the susceptibility of CD34– stem cells to retroviral transduction and their subsequent engraftment potential is a matter of some priority, since this population is currently being discarded in the overwhelming majority of transduction protocols. The importance of purification of stem cells prior to transduction is evident in situations where drug resistance genes, such as MDR or mDHFR, are used to transfect the BM or mPB cells of cancer patients. In this situation, HSC selection minimizes the possibility that malignant cells will contaminate the population to be transduced. CD34+ cell selection can deplete contaminating breast cancer cells by two-to-four logs, but is rarely 100% efficient. The cell culture conditions used for cycle activation and transduction of HSC are also incompatible with proliferation and viability of tumor cells, so there is a further one-to-two log depletion of tumor cells during this phase. Ultimately, positive selection strategies to remove tumor cells may be required, for example with anti-epithelial antibodies in the case of breast cancer.

Retroviral Receptors and Envelopes

MLV and lentiviral vectors may be pseudotyped into different viral envelopes, depending on the retroviral packaging system employed. The retroviral envelope binds to specific cell surface receptors and thereby determines the viral host range [reviewed in ref. 36]. Currently, the envelopes most commonly employed for transduction of human HSC are derived from either amphotropic MoMLV [37, 38], gibbon ape leukemia virus (GALV) [39], or vesicular stomatitis virus-G (VSV-G) [40]. The question of the most suitable envelope for infection of HSC has been addressed in several investigations that compare either expression of retroviral receptors on progenitor cells or compare transduction efficiency of vectors pseudotyped in different envelopes.

Quantification of expression of the amphotropic MLV receptor (Ram-1) in human hematopoietic progenitors was first reported by Orlic et al. [41]. By RT-PCR, expression of Ram-1 mRNA was low in BM-derived CD34+CD38− cells. Kiem et al. [42] reported higher levels of the GALV receptor (Glvr-1) than Ram-1 in human CD34+/38− cells and baboon-derived CD34+ cells. However, in a more recent study, Orlic et al. [43], showed no significant difference in the level of expression of Ram-1 and Glvr-1 in freshly isolated human BM, CB and FL-derived CD34+CD38− cells, although expression of both receptors appeared to be slightly higher on CB and FL than on adult BM cells. The latter study also demonstrated that amphotropic and GALV receptor mRNA levels were three-fold higher on BM CD34+CD38+ than on CD38− cells. CD34+CD38− cells bound <10% of the amount of virus that bound to CD34+CD38+ cells [44]. These results are consistent with the observation that primitive cells, such as CD38− cells, are more difficult to transduce than more mature progenitors. Comparison of fresh and frozen CB cells has shown that Ram-1 but not Glvr-1 mRNA expression was four fold higher on frozen and thawed cells than fresh CD38+ cells and twelve fold higher on frozen, thawed CD34+CD38− cells [43]. This is a particularly fortuitous observation, in view of the potential use of cryopreserved cells for clinical gene therapy, and further investigation of the mechanisms involved, for example the role of DMSO in receptor upregulation, is warranted. Addition of interleukin-3 (IL-3), IL-6, and c-kit ligand (KL) (also referred to as Steel factor and stem cell factor) to cultures of BM CD34+ cells was also shown to increase amphotropic retrovirus binding [44].

To compare transduction of HSC by GALV pseudotyped MLV-based (GALV/MLV) vectors and amphotropic MLV (ampho/MLV) vectors, Kiem et al. [41, 45], performed competitive repopulation assays in baboons. The GALV/MLV vector appeared to more efficiently transduce primary CFC and contributed a greater percentage of gene-marked cells to the peripheral blood and BM than the ampho/MLV vector of equal titer. However, engraftment of gene-marked cells was low with both vectors; less than 5% of transduced cells were detected in the peripheral blood and BM. Also, it is unclear whether this model is relevant to humans, as it is conceivable that GALV-pseudotyped vectors transduce primate cells more efficiently than human hematopoietic cells. However, a recent study by independent investigators also indicated that GALV-pseudotyped vectors may be preferable to amphotropic vectors for transduction of human CB-derived LTC-IC [46]. Marandin et al. [46] compared transduction by GALV and amphotropic vectors derived from both murine and human packaging cell lines. The human cell-derived amphotropic and GALV vectors more readily transduced LTC-IC than a murine cell-derived amphotropic vector. When titers of the human cell-derived vectors were equilibrated, the GALV packaged vector transduced LTC-IC more efficiently than the amphotropic vector. For gene therapy purposes, MLV vectors produced by human cell lines may be critical because vectors produced by murine cell lines are inactivated by human complement [47–49].

VSV-G pseudotyped vectors are potentially advantageous because they can be concentrated to very high titers by ultracentrifugation [40]. VSV-G packaged vectors have been shown to transduce human hematopoietic progenitors, including LTC-IC [50–52]. In the study reported by Case et al. [52], VSV-G pseudotyped MLV vectors (VSV-G/MLV) more efficiently transduced progenitor cells than GALV/MLV when the titers of the two vectors were equivalent. In an investigation comparing VSV-G and amphotropic packaged vectors, VSV-G/MLV vectors transduced LTC-IC more efficiently than ampho/MLV, even though transduction of primary CFC was similar with the two vectors [51]. It was not clear however, from the latter report, whether the titers of the two vectors were equivalent. One complication in the application of VSV-G packaging systems is that constitutive expression of the VSV-G envelope protein is cytotoxic. In addition, VSV-G envelopes appear to be toxic to primary cells, including CFC [53]. Due to the problem of toxicity, stable VSV-G producer cell lines have not been developed. Instead, investigators have employed transient or inducible packaging systems for production of VSV-G pseudotyped vectors [54, 55].

Human cells may also be transduced by ecotropic MLV if the target cell is engineered to express ecotropic receptors from an adenoviral or adeno-associated viral vector [56, 57]. Adenoviral vectors transduce functionally primitive progenitor cells with relative efficiency [58]. Transduction of human hematopoietic cells via an adenovirally expressed ecotropic receptor may exploit the benefits of both adenoviral and retroviral infections; the adenoviral vector would enable transduction of primitive subsets of cells and the retroviral vector would be stably expressed in daughter cells.

To date, there is no consensus on the best choice of envelope for transduction of human HSC. Further investigations will be required to develop novel transduction systems and to unequivocally determine which viral envelope is most suitable for human HSC. Gene marking studies in transplanted human patients may be the only way to make these determinations.

Marker Systems and Assays Used to Evaluate Transduction Efficiency

Semiquantitative PCR methods and individual colony PCR analysis has been used extensively for measurement of efficiency of viral integration. The percent of primary CFU-GM, BFU-E, or CFU-GEMM, that are PCR positive is usually high in populations of CD34+ cells plated shortly after transduction, however this does not provide an index of transduction of HSC. As a surrogate assay for stem cells, the LTC-IC assay is most frequently used. The LTC-IC assay requires transduced cells to be co-cultured for at least 5 weeks on a preformed murine or human BM stroma before the adherent layer and suspension cells are plated for secondary colony formation. PCR analyses for proviral integrants have generally demonstrated a higher frequency of transduced primary colonies than secondary colonies, which reflects the difficulty of transducing more primitive cells. The drawback of the PCR evaluation is its failure to evaluate the efficiency of transcription and translation of the transgene. In situations where gene silencing occurs, for example, with MoMLV LTR, PCR positive secondary colonies overestimate the percentage of transgene-expressing cells. In very sensitive PCRs there is also the problem of contamination of colonies by background PCR+ non colony cells or neighboring PCR+ colonies, this is particularly a problem in the secondary colony assay where relatively high numbers of cells are plated and persisting macrophages derived from CFU-GM, which are transduced with high efficiency, may contaminate the colonies. A control using intercolony areas is useful but does not eliminate the problem since there may be differential survival of background cells within a colony environment.

When the neomycin resistance (NeoR) gene is employed, selection for G418-resistant colonies is often used as a measure of transduction efficiency. In primary and secondary CFC assays, progenitor survival is exponentially related to G418 concentration [59]. It is therefore theoretically impossible to get zero survival of CFC. In practice, different investigators use quite widely different concentrations of G418 (0.1–3.0 mg/ml), with the majority of investigators using 1.0–1.5 mg. Furthermore, there can be considerable batch variation in activity and degree of hydration of the drug. Most important, G418 resistance is rendered an unreliable readout of transduction by the existence of highly resistant non-transduced progenitors and a documented increase in resistance of non-transduced progenitors following cytokine priming [59]. The dose of G418 needed to give 1% survival of normal BM BFU-E was reported to be 1.22 mg/ml whereas equivalent survival of CFU-GM required 2.4–2.7 mg/ml [60]. Chinswangwatanakul et al. [59] also showed that the cell source as well as the type of progenitor influenced G418 sensitivity; CB-derived BFU-E required a higher concentration of G418 to achieve 99% inhibition (2.4 mg/ml) than did BM-derived BFU-E. A 3 day stimulation of CD34+ cells with cytokines rendered both CB and BM CFU-GM and BFU-E more resistant to inhibition, with up to twice the concentration of G418 needed to give 99% inhibition. Thus, the validity of G418 resistance requires that the control population be manipulated as closely as possible to the transduced population, and different G418 doses be used for separate evaluation of CFU-GM and BFU-E. Also, PCR amplification of vector-derived sequences in resistant colonies may be used to validate transduction efficiency determinations made as a percentage of G418-resistant colonies.

The β-galactosidase gene has been used as a marker in a number of studies of CD34+ transduction, with up to 36% of CFC positive by X-gal staining [46, 60, 61]. Introduction of genes encoding marker proteins allows for viable cell sorting and positive selection of transduced cells, as well as ready monitoring of transduced cells following in vivo transfer. For example, placental alkaline phosphatase has been used as a marker in canine BM transplantation experiments, with 5–10% of PB leukocytes expressing the marker a month after transplantation [62]. Transduction with the gene for the murine heat-stable antigen (HSA) produced expression in 27% of CD34+ cells and 7% of LTC-IC [63]. A marker protein that will not evoke an immune response is desirable if positive selection and marker studies are to be extended to patient populations. Mutated forms of the human low affinity p75 NGFR are useful markers that can be expressed at high levels on the cell membrane, but are rendered unable to transmit signal transduction or bind ligand [32, 64–67]. Under optimal transduction conditions, NGFR expression was detected on 30–50% of CD34+ cells [32, 66, 67], including Thy-1+, Thy-1– [66] and HLA-DR– populations [65], with comparable expression on primary CFC. NGFR+ CD34+ cells, and CFC can be detected after 5 weeks of culture, but at a much lower level, unless transduced cells are enriched by positive selection at the outset [65]. Another useful marker is the green fluorescent protein (GFP) which displays autonomous fluorescence and thus eliminates the need for antibody or cytochemical staining. In animal studies, GFP was used as a transduction marker for prolonged periods in vivo without toxicity and without evoking an immune response [68]. GFP expression also apears to be non toxic in human progenitor cells. Independent investigators utilizing retroviral vectors expressing GFP have reported transduction of 25–40% of human CD34+ cells, up to 40% of CFC and 2–8% of LTC-IC without apparent toxicity [69–71].

Cytokine Priming Strategies for Stem Cell Cycle Activation

The requirement for cells to be in cycle and undergo mitosis for successful retroviral integration to take place [72, 73] has necessitated development of various strategies to activate normally quiescent stem cells into cycle. In one gene-marking trial performed in young patients, relatively efficient transduction of long-term repopulating stem cells was achieved without cytokine pre-treatment and with a simple 6-hour exposure of cells to virus [34]. However, in this case, BM cells were obtained 2–6 weeks after high-dose chemotherapy and as a consequence of the chemotherapy, a significant fraction of HSC would have been proliferating at the time of retroviral infection. The marker gene was present and expressed in all hematopoietic lineages in 5/5 of these patients after 1 year. The latter study was exceptional since, without cytokine priming, transduction of progenitors and primitive cells such as LTC-IC, cobblestone area forming cells (CAFC), or SRC from normal adult BM, CB and G-CSF/cytoxan mPB is negligible. In one representative study, retroviral infection of heavily pretreated adult BM CD34+ cells for 6 h without hematopoietic growth factors or stromal support resulted in only a low level of gene expression (0.01–1.00% positive cells) in transplanted patients at 6 months [74].

For unstimulated cells, the fraction of CD34+ cells progressing through S-phase is 1–2% in CB [27, 75] and 0–8% in mPB [76]. Cell cycle activation of CD34+ cells is readily induced by 48- to 72-hour exposure of the cells to a combination of synergistic factor(s) such as KL and Flk-2/Flt-3 ligand (Flk-2L) with a hematopoietic growth factor(s) such as IL-3, GM-CSF, G-CSF, IL-11, IL-1, IL-6 or thrombopoietin (TPO). Cytokine combinations most appropriate for priming cells for retroviral transduction should be capable of stimulating cycling of primitive CD34+ subpopulations, such as CD38– cells and functionally primitive LTC-IC/CAFC. Also, the progeny of such cycling cells should retain stem cell potential and in vivo long-term repopulating potential.

Many groups have attempted to determine optimum cytokine combinations for ex vivo expansion of stem cells in the absence of stromal support, using various combinations of the above cytokines. Our own studies with G-CSF+cytoxan mPB CD34+ cells from patients with ovarian cancer or heavily pretreated breast cancer indicated that, using a combination of KL+G-CSF+IL-1 or IL-6 and IL-3 in a closed, gas-permeable bag system with autologous plasma, a mean of 100- to 150-fold expansion of CFU-GM could be obtained in 10–12 days with retention of input numbers of LTC-IC [77]. The same cytokine combination applied to CB cells produced a 200 to 1,000-fold expansion of progenitors and a 15-fold expansion of LTC-IC in 14 days [27]. More recently, a number of studies have shown that Flk-2L synergizes with IL-3, IL-6, and G-CSF or TPO, to promote greater LTC-IC expansion than combinations with KL [28, 78–81].

Retroviral transduction protocols generally involve a total ex vivo culture duration of 3–5 days which includes a 24- to 72-hour cytokine priming phase, followed by a further 24–72 h in the presence of cytokine and virus. In the case of marker studies requiring protein expression (e.g. GFP or NGFR), a further 48–72 h may be added to ensure optimal transgene expression. In the last 5 years, most cytokine combinations have included KL at concentrations of 10–100 ng/ml, together with IL-3 and IL-6 [33, 43, 64, 82–84], or IL-3 and IL-1 [77]. These cytokine combinations generally resulted in reasonable transduction of CD34+ cells (10–30%) and primary CFC (10–50%), but relatively poor transduction of LTC-IC (week 5 CFC), and low or absent engraftment in immunodeficient mice [81, 85], and humans [83, 86]. It appears that, in these studies, the CD34+ cells that were most rapidly activated into cycle, and therefore transduced, had limited ability to produce assayable progenitors in long-term assays. The cells that were responsible for maintenance of long-term hematopoiesis were probably resistant to the conditions employed for cytokine stimulation and were therefore refractory to transduction.

In order to evaluate cytokine-driven cell cycle activation of primitive cells, Veena et al. [87] used a fluorescent membrane dye to isolate CD34+ cells that were refractory to initial cytokine stimulation. The selected quiescent cells were evaluated for ability to generate long-term progeny as well as susceptibility to further cytokine priming and retroviral transduction. The initially cytokine-resistant population was shown to include long-term repopulating cells that were transducible following further cytokine stimulation. It is possible that 24- to 48-hour cytokine prestimulation moves HSC from deep dormancy to a stage when they are primed to enter cell cycle. Consistent with this possibility, Boezeman et al. [88] noted a 2.6- to 3.1-day delay in initiation of proliferation of the most primitive CD34+ subpopulation when exposed to cytokines in vitro. The kinetics of cell cycle entry and exit of primitive hematopoietic cells in culture is particularly complex. Division of CD34+CD38– cells was directly observed by fluorescence labeling with PKH-26 [90]. In the presence of KL, IL-3, IL-6, GM-CSF and Epo, 17% of these primitive cells from adult BM, 31% of CB and 57% FL clonally proliferated. This study demonstrated that 36–48 h were required for completion of the first mitosis. Each subsequent division took only 12 h; however, consistently $\sim 40\%$ of all initial cell divisions were asymmetric – one daughter cell became quiescent while the other proliferated exponentially. The ratio of asymmetric divisions to symmetric divisions did not depend upon the cell source, although the percent of cells undergoing asymmetric division decreased with ontogenic age. The implication of these observations to retroviral transduction strategies is considerable. If, for example, the quiescent daughter cell retained stem cell function while the proliferating partner underwent differentiation, prolonging the duration of virus exposure would not necessarily increase transduction of primitive cells. Strategies designed to reactivate quiescent daughter cells, or to block reentry into quiescence (for example by anti-TGFβ treatment) will need to be employed. Extending the duration of exposure of cells to the above cytokine combinations and virus exhibits the law of diminishing returns since there is progressive loss of transduced cells capable of long-term repopulation in immunodeficient mice. Dao et al. [84, 85, 90] compared 72 h of cytokine priming (IL-3, IL-6, KL) with virus addition at day 1–3, to priming extended to 7 days with virus addition on days 5–7. Engraftment of immunodeficient mice with transduced cells was 1.85% following the shorter duration of cytokine priming, versus 0.08% with transduced cells subjected to extended priming.

In efforts to improve upon transductions performed with the basic cytokine combinations discussed above, additional cytokines have been utilized, including G-CSF [91], Epo [67] and basic fibroblast growth factor (βFGF) [92]. The latter significantly increased the generation of G418-resistant CFU-GM (38%) and CFU-GEMM (35%). Flk-2L has also been extensively employed, generally in combination with KL, IL-3 and IL-6. The addition of Flk-2L to standard cytokine cocktails promoted transduction of BM CD34+ cells, including CD38– populations that subsequently expressed the transgene in LTC-IC or immunodeficient mice. In the latter investigations, transduction with a cytokine combination that included Flk-2L resulted in higher transduction than achieved with the same cytokine combination plus BM stromal cells [84, 85]. Flk-2L was also effective in improving transduction efficiency when combined with IL-3 alone or IL-3, IL-6, and KL [93]. The combination IL-3, KL, Flk-2L was effective for transducing LTC-IC within CD34+ primitive subpopulations, including CD38lo and HLA-DRlo [94]. Efficient transduction of LTC-IC (36%) was also obtained following priming with Flk-2L and IL-3, IL-6 and KL [71]. In another independent study, the use of Flk-2L with a number of additional cytokines (IL-3, IL-6, GM-CSF, TPO, KL) more efficiently primed CD34+ cells for transduction than IL-3, IL-6 and KL with 21% CD34+ cells, 36% of CFC and 8% of LTC-IC transduced [69]. The latter combination also enabled optimal transduction of CD34+CD38– derived primary CFC (60%) and LTC-IC [46].

While direct comparisons of cytokine cocktails are difficult because of variables other than cytokine combinations, doses and duration (e.g. cell source, virus envelope, method of transduction and particularly viral titer), the majority of studies point to the value of addition of Flk-2L for transduction of early cells, particularly LTC-IC. However, even with Flk-2L, prolonged culture is deleterious to preservation of cells with in vivo long-term reconstituting potential. Direct evidence for the negative effect of prolonged culture was provided in a series of competitive repopulation studies performed in rhesus monkeys [95]. In those studies, CD34+ cells were primed and expanded with IL-3, IL-6, KL +/– Flk-2L +/– stromal support, with cryopreservation of one aliquot of cells after 4 days in culture and continued expansion of another aliquot for a further 10 days. Four-day and 14-day culture populations were differentially gene marked and combined for transplantation into autologous recipients. The contribution to long-term hematopoiesis by different gene-marked populations was evaluated. This informative study revealed that, despite 5- to 13-fold higher numbers of cells and CFC in the 14-day expanded population, there was greater short- and long-term marking with the short-term cultured cells. The 4-day culture with addition of Flk-2L and the use of stroma increased engraftment to a clinically relevant range (10–20% marked cells), compared to IL-3, IL-6 and KL alone (0.01% marked cells).

It is a paradox that transduction of the most primitive populations, including LTC-IC, increases with duration of cytokine exposure to cytokine combinations that include Flk-2L and is also associated with progressive expansion of LTC-IC numbers (with time) and progressive loss of in vivo repopulation capacity [96]. Autologous transplantation of ex-vivo-expanded mPB CD34+ cells in myelosuppressed individuals showed comparable time and extent of platelet and neutrophil recovery to that achieved with 10-fold more non expanded cells [97]. However, recent reports indicate that long-term engraftment in vigorously myeloablated individuals may be compromised by transplantation with ex-vivo-expanded PB CD34+ cells that otherwise include optimal numbers of CD34+ cells and LTC-IC [98].

There are conflicting reports on the effects of brief cytokine exposure on BM repopulating ability. Tavassoli et al. [99] reported that 2- to 3-hour preincubation of mouse BM with IL-3 enhanced repopulating ability, possibly due to upregulation of homing receptors. In contrast, van der Loo and Ploemacher [100] found that a similar preincubation with IL-3 or IL-3+IL-12+KL led to sustained decrease in BM and spleen seeding of both early and late CAFC and a reduction in day-12 CFU-S seeding from 11.4 to 7.3%, together with a decrease in long-term repopulation. It is clear that IL-3 addition can reduce long-term repopulating ability of cultured BM cells [101–103]. This negative regulation by IL-3 appears to be mediated both by the common receptor signaling subunit βc and the additional IL-3 signaling protein β IL-3 which is specific to IL-3 and is found in the mouse but not in man [103]. One explanation for the discrepancy between the generation of LTC-IC in vitro, higher transduction efficiencies, and the lack of long-term in vivo engraftment of marked cells is that LTC-IC and CAFC assays do not measure 'true' HSC responsible for in vivo long-term repopulation. In vivo assays for stem cells are more rigorous due to the requirement for retention of stem cell homing capacity and chemotactic responsiveness. Engraftment of human hematopoietic cells within the BM of immunodeficient mice after 5–6 weeks has been used to evaluate the in vivo repopulating capacity of ex-vivo-expanded CD34+ populations. In a number of studies of ex vivo expansion of CD34+ cells, the number of LTC-IC was at least equivalent, if not greater, than in the input population after 10–14 days, yet NOD-SCID repopulating capacity was lost [96, 104]. Mobest et al. [96] reported that when PB CD34+ cells were cultured in serum-free medium with Flk-2L (300 ng/ml), KL and IL-3 (100 ng/ml), there was a 10- to 50-fold expansion of CFC by 7–10 days and a 1- to 3-fold expansion of LTC-IC, peaking at 6–8 days. However, during this time, there was a progressive loss of NOD-SCID mouse lymphomyeloid engraftment (13% long-term human engraftment with non expanded CD34+ cells, 11% engraftment with 2-day expanded cells, 8% engraftment with 4-day expanded cells and no engraftment with 7-day expanded cells). In some studies, in vivo repopulating capacity of expanded CB cells increased at 4 days by 2- to 4-fold, but was lost by 9 days [105], or increased by 2-fold at 5–8 days [106]. In stromal coculture, a 6-fold loss of repopulating capacity was reported with CB cells over 14 days, despite LTC-IC being higher after expansion [104]. CB cells expanded for 6 days with IL-3 and KL failed to engraft when injected intravenously in NOD/SCID mice (but did so when engrafted intraperitoneally) [107]. Following culture for 14 days with KL, Flk-2L, IL-6, Epo either with or without IL-3, BM CD34+ cells injected into fetal lambs produced marrow engraftment, but in contrast to non cultured cells, the expanded cells did not sustain long-term post-natal engraftment [108]. Together, these observations suggest that ex-vivo-expanded stem cells have acquired defects in marrow-homing capacity, which prevents localization to the BM, possibly resulting in their clearance and destruction in non hematopoietic tissue. Specifically, IL-3 appears to be instrumental toward this effect. Adult CD34+ populations expanded with IL-3-containing cytokine cocktails showed a major loss of

NOD/SCID repopulating capacity at 6–7 days with BM [109] or mPB [96]. In contrast, the cytokine combination TPO, KL and Flk-2L for 6 days resulted in an increase in CAFC numbers and a capacity for NOD/SCID repopulation similar to fresh BM [106, 107, 109]. This result parallels the observation that a similar cytokine combination expanded CAFC and LTC-IC for 7 and 14 days without loss of chemotactic responsiveness, and that these features were compromised by addition of IL-3 [110].

Transforming Growth Factor-β and HSC Transduction

Transforming growth factor-β (TGF-β) is a major physiological negative regulator of stem cell proliferation. TGF-β appears to counteract the proliferative stimuli provided by KL and Flk-2L in synergy with other growth factors, and either maintains cells in G_0 or returns them to that state [111, 112]. The effects of TGF-β are partially elicited by inhibition of c-kit and Flk-2 mRNA production, with resulting downmodulation of these receptors on HSC [113, 114]. In BM stromal coculture, TGF-β production by stromal cells places HSC within the adherent layer in a noncycling state that can be overcome by addition of anti-TGF-β serum to the cultures. Autocrine production of TGF-β by CD34+ cells has been documented, and addition of anti-TGF-β serum or antisense oligonucleotides to cytokine-stimulated CD34+ cultures increased the fraction of cells entering cell cycle and enhanced expansion of the population [111, 115].

Anti-TGF-β serum has been shown to enhance retroviral transduction of CD34+ cells, LTC-IC, and NOD/SCID repopulating cells [116]. Schilz et al. [117] primed BM or CB CD34+ cells with KL, FL and IL-6, and anti-TGF-β in an optimized transduction protocol, to achieve transduction of up to 97.4% of CD34+ cells and 5.6–56.6% SRCs. Production of TGF-β in cultures may down modulate viral receptors or the receptors for adhesion molecules that are required for stem cell homing to marrow. In this context, we have observed that brief (24 h) exposure of CD34+ cells to TGF-β significantly reduces the transendothelial chemotactic response of CAFC and LTC-IC to SDF-1, possibly as a result of downmodulation of the chemokine receptor, CXCR4.

Inactivation of TGF-β may be particularly useful for transduction of CD34+ cells from HIV-infected individuals. Potential gene therapy for AIDS has promoted investigators to compare transduction of CD34+ cells from HIV+ patients versus cells from normal or from cancer patients. Kearns et al. [117] showed that BM CD34+ cells from both HIV+ and HIV− children were equally susceptible to retroviral transfection. However, a significant transduction defect was seen in the HIV+ adults [51]. Increased TGF-β levels have

Table 3. Factors determining stable transduction of human HSC capable of long-term multilineage repopulation

---

Retroviral half-life (5–8 h at 37 °C)
Distance the virus has to travel before interacting with cells (rate governed by Brownian motion)
Repulsive steric and electrostatic forces at cell surface
Viral concentration-multiplicity of infection
Number of viral receptors per cell and binding affinity (amphotropic, GALV, VSV-G)
Efficiency of internalization of virion core into cell cytoplasm
Rate of intracellular decay of virus (half-life 5.4–7.5 h)
Efficiency of reverse transcription of viral RNA into DNA in the core structure
Entry of double strand DNA as a nucleoprotein complex into nucleus
Integration of retroviral DNA into the stem cell genome
Integration of MLV-based vectors requires that stem undergo mitosis
Cytokine combinations that activate stem cells into cycle may or may not maintain stem cell phenotype
Ability of transduced stem cells to engraft may be compromised by ex vivo cycle activation and expansion
Transgene expression in stem cells may be downmodulated e.g. by methylation
Transgene proteins may evoke a host immune response
Clonal expansion of limited numbers of transduced stem cells may lead to proliferative senescence

---

been reported in HIV-infected individuals, and recent studies have shown that HIV virus may induce TGF-β production by CD34+ cells [118], presumably by signaling through the viral coreceptor CXCR4, which is expressed at high levels on CD34+ cells and HSC. Thus, TGF-β may indirectly inhibit proliferation and block retroviral transduction of CD34+ cells from HIV-infected adults.

Retroviral Transduction Strategies

Interaction of the retrovirus with the target cell and subsequent viral internalization, reverse transcription and integration into the target cell genome are physicochemical factors that determine the efficiency of the transduction process (table 3). Retroviruses are colloidal particles which, in passive transduction systems, come into proximity with target cells by Brownian motion. In static systems, the gravitational settling velocity of the virus is such that the average retrovirus with a half-life of 5–8 h at 37 °C can traverse only 480–610 μm within a half-life [119]. Thus, in a static culture system of 2–5 mm depth, the majority of viral particles will be unable to reach the cells within their normal half-life. These physicochemical limitations can be overcome by using centrifugation, flow-through strategies, or immobilizing supports to more rapidly approximate virus to target cells.

*Conventional Static Supernatant Transduction and
Producer Cell Coculture*

Coculture with viral producer cells is always more efficient than simple addition of viral supernatant, since the producer cells provide a continuous source of fresh virus and a site for co-localization of the retrovirus and target cells. However in cocultures, there is a very significant loss of SRC [120] and LTC-IC [32], largely due to depletion of HSC by adhesion. Efforts must therefore be made to recover primitive cells that are strongly adherent to producer cells. Regulatory issues would also arise if coculture were used for clinical gene therapy.

A conventional supernatant infection protocol was successfully used to transfect long-term repopulating cells in one pediatric clinical trial [34], but not in an adult study [74]. In the pediatric trial, fresh or frozen, thawed virus was added to BM cells for 6 h. Supernatant transduction has been evaluated in a series of studies using virus supernatants at m.o.i. ranging from 1 to 10 following variable durations of cytokine priming. A single addition of virus, or two additions over 48 h, led to 16–25% CFC transduction [28, 51]. Three additions over 72 h gave 21–37% CFC transduction [82, 84, 85, 90, 92, 121, 122] and four additions over 96 h gave 58–59% transduction [31]. In a direct comparison of the impact of number of additions of virus on cumulative transduction of CFC, Marandin et al. [46] observed transduction of 14% of CFC with one round of virus, rising to 80% of CFC with eight rounds over 72 h. Transduction in supernatant and coculture infections is influenced by a number of additional factors, including the producer cells (discussed above) and the use of polycations. Polycations promote retroviral transduction by neutralizing electrostatic repulsion between opposing bilayers of the virus and cell surface [123]. Polybrene and protamine have been used at concentrations of 4–8 µg/ml in retroviral infections. However, protamine is preferable for clinical studies in view of accumulated evidence that polybrene has a dose and time-dependent inhibitory influence on the most primitive cells responsible for long-term hematopoiesis [76, 124].

Studies with serum-free conditions have also been undertaken, because the presence of fetal calf serum in the viral supernatant and in the transduction medium is of concern in clinical protocols. Efficient transduction can be obtained without serum [124]. Indeed, better transduction was observed without serum in one study where viral titers were comparable [67, 125]. However, the viral titer is generally lowered by production under serum-free conditions

[124]. Recently, a serum-free formulation with defined pharmaceutical grade reagents was developed. This formulation is apparently not detrimental to production of high titers of virus [126].

Centrifugation Transduction – 'Spinoculation'

Centrifugation increases the rate of virus-cell association and decreases the rate of dissociation, resulting in overall increase in probability of virus uptake. Transduction was reported to be directly proportional to the time of centrifugation, centrifugal force (up to 10,000 g), and inversely proportional to target cell numbers [127, 128]. Centrifugation for 1 h at 10,000 g and at 32 °C rather than 37 °C resulted in transduction of 95% of CFU-GM as determined by PCR. Centrifugation of a GALV-pseudotyped vector expressing NGFR with human CD34+ cells for 90 min at 2,500 rpm at 32 °C, resulted in >70% of CD34+ cells transduced, which could be increased to 98% by combining 'spinoculation' with fibronectin fragment colocalization of virus and target cell [116]. Abe et al. [31] centrifuged CD34+ cells for 4 h at 1,800 g at 32–35 °C at a cell-to-virus ratio of 1:1, to obtain >30% NGFR+ cells. Campain et al. [30] compared passive or static amphotropic virus infection of CD34+ cells with 3 round of spinoculation for 2 h at 2,500 g. In contrast to other investigators, the latter failed to see a significant difference in transduction efficiency; spinoculation resulted in 19.4±8.5% transduction, whereas 16.2±4.8% cells were transduced in static cultures. However, the spinoculation procedure took 6 h, versus 48 h with less manipulation. Lower temperatures are used in spinoculations so that the cultures do not become too warm (>37 °C) due to the centrifugal forces. Lower temperatures are also advantageous for static transductions, because the rate of virus inactivation is reduced. It has been shown in primate BM transductions that virus exposure at 33 °C rather than 37 °C increased transduction 1.6±0.4-fold [129].

Flow-Through Transduction

Virus may be actively flowed over cells to increase the frequency of virus-cell interaction. Fluid flow may be achieved by agitation or mechanical mixing/stirring of the cells with virus. However, the micro-hydrodynamics are such that laminar hydrodynamic boundary layers form close to solid surfaces and are of such a thickness that they may preclude virus penetration. Even with bulk fluid agitation, the final encounter of the virus and cell is governed by Brownian motion [119]. This limitation can be overcome by first seeding the target cells onto porous surfaces and flowing a virus solution directly through a layer of target cells [130, 131]. Using model cell lines in flow-through systems, it was shown that gene transfer rates can be increased by an order of magnitude using the same concentration of infection medium. Using a collagen-coated membrane in a flow-through system, high transduction rates were obtained, even in the absence of polycations [119, 130]. Transduction efficiency increased with the flow rate until a plateau was reached at average flow velocities that exceeded 0.3 cm/h for flow durations of 3–4 h [130]. A correlation was found between optimal time for maximal gene transfer using flow-through transductions and the optimal time for maximal virus activity on the membrane suggesting that the membrane adsorption capacity for virus determined the amount of gene transfer that could occur [130, 131].

Continuous perfusion systems have been used to improve transduction of CD34+ cells. Primary BM stroma [132] or engineered stromal cell lines [133] were seeded with CD34+ cells and transduction was significantly improved by continuous flow perfusion compared to passive addition of virus supernatant. Bertolini et al. [133] perfused a virus expressing MDR for 7 days in a capillary perfusion system and obtained 74–79% P glycoprotein positive (Pgp+) CD34+ cells and 88% taxol-resistant LTC-IC in the perfusion system whereas in the static system, only 42–49% of CD34+ cells were Pgp+ and 68% of LTC-IC were taxol-resistant. Eipers et al. [132] reported that 100% of CFU-GM were PCR+ after 1 week in a perfusion system, and 33% of 7-week CFU-GM were vector-positive by PCR. However, perfusion of a layer of cells on a permeable membrane was more effective for adherent cells such as fibroblasts than for CD34+ cells, due to the latter's greater fragility and resultant cell loss in the perfusion procedure [66]. In the course of evaluating a perfusion system, Hutchings et al. [66] observed that it was possible to achieve high transduction efficiencies by concentrating retroviral supernatant by flow-through 0.2-µm pore membranes in the presence of polybrene, then subsequently adding CD34+ cells and culturing overnight on the filter surface. In the latter study, the use of the filter system improved transduction of CD34+ cells from 4% with an m.o.i. of 10 and no cytokines in a conventional transduction, to 17% with use of the filter system and 39% with cytokine prestimulation and culture on a filter. With an m.o.i. of 40 and no cytokine stimulation, 46% of the cells were transduced in the filter system.

Fibronectin

Several independent investigators have demonstrated that retroviral gene transfer to human hematopoietic progenitors may be enhanced by using plates coated with human fibronectin or fibronectin fragments [18, 134, 135]. Fibronectin is a BM extracellular matrix protein that mediates adhesion of human hematopoietic progenitor cells through cell surface proteoglycans and the β1 integrins, VLA4 and VLA5. The cell-binding domain, heparin-binding domain, and CS1 sequence of the fibronectin molecule appear to be critical for hematopoietic cell adhesion [135]. Hanenberg et al. [136]

demonstrated that amphotrophic MLV particles also bind to the heparin binding domain of fibronectin fragments. These observations lead to the proposal that fibronectin facilitates retroviral infection of CD34+ cells by co-localizing viral particles with the progenitor cells [135, 136]. Initial studies also suggested that the use of fibronectin in gene transfer protocols overrides the necessity for polycations such as polybrene and protamine. However, a recent report clearly indicates that the inclusion of polycations in addition to fibronectin fragments provides a further increment in transduction of primitive hematopoietic cells [44].

A recombinant human fibronectin fragment (CH-296 or Retronectin), which contains the critical domains necessary for both cell adhesion and virus binding [136], appears to better enhance transduction of CD34+ cells than full-length fibronectin [20]. However, CH-296 does not necessarily enhance transduction of the more primitive subsets of cells. In gene transfer experiments utilizing CH-296 fragments, Hanenberg et al. [137] reported transduction of only 3–17% of CD34+/CD38– cells, despite transduction of up to 68% of CD34+/CD38+ cells. SRC also appear to be more refractory to infection on CH-296 fragments; Larochelle et al. [120] reported up to 80% transduction of primary CFC, but less than 3% transduction of SRC cells. Schilz et al. [116] reported more encouraging transduction of SRC by combining Retronectin with the spinoculation technique. Cytokines which stimulate the more primitive progenitor cells appear to be necessary for CH-296-mediated transduction of repopulating cells. In a recent study, Kiem et al. [45] demonstrated that CH-296 offers no significant advantage over coculture when baboon CD34+ cells are cultured in KL, IL-3, and IL-6. However, when the cells were prestimulated and cultured in KL, IL-6, TPO and Flk-2L, significantly more long-term repopulating cells were transduced in CH-296-assisted infections than in cocultures. The latter results reiterate the importance of early-acting cytokines and demonstrate that both HSC proliferation and cell-virus interactions were unfavorable for efficient gene transfer in previously employed protocols.

Engraftment of Transduced Human Pluripotent Stem Cells in Immunodeficient Mice

Three immunodeficient murine model systems have been utilized to evaluate gene transfer into human hematopoietic precursors, with long-term (3–12 months) multilineage engraftment potential. Two of these, the NOD/SCID and the beige/nude/X-linked immunodeficient (bnx) mice support human hematopoiesis in the murine BM environment, whereas the third model, the

SCID-hu model, supports human hematopoiesis in grafted fetal human bone and lymphopoiesis in grafted fetal human thymus. The advantage of the SCID-hu system is that transduced cells are injected directly into a human marrow or thymic microenvironment, thus overcoming problems associated with the low seeding efficiency of intravenously injected human cells ($<1\%$ of human progenitor cells home to the BM), or species barriers involving certain cytokines, chemokines or adhesion molecules. The result is a high level of durable donor engraftment in BM and donor T cell differentiation in the thymus. Cheng et al. [33] used this model to assess engraftment of mPB CD34+ cells transduced with a vector driven by the murine stem cell virus promoter and carrying the NGFR marker gene. Following two cycles of 'spinoculation' with IL-3/IL-6/KL, 13% of CFC and $\sim 3\%$ of LTC-IC were vector-positive by FACS analysis. Intrabone injection resulted in engraftment of 11–14% of donor cells at 9 weeks and, of these, 28–45% expressed NGFR. Similarly, 28–45% of human CD19 B cells were NGFR+. In another study utilizing SCID-hu mice, FL CD34+ cells were transduced with an amphotropic vector expressing a β-galactosidase marker. Ten percent of the cells injected into SCID-hu mice were β-Gal+, and 4 months later, low levels ($<0.2\%$) of marked cells were detected in BM and thymus, with 5% of hu-CFC also marked [61].

In the bnx model, long-term (>12 months) stable engraftment of human hematopoietic and T lymphoid (but not B lymphoid) cells may be achieved, albeit at low levels. However, engraftment may be improved by coinjection of stromal cells engineered to express human IL-3 [82, 84, 85, 90, 122, 138]. In a series of studies by the latter investigators, BM and mPB CD34+ cells were transduced with vector carrying NeoR by three 24-hour cycles of virus supernatant addition in the presence of IL-3/IL-6/KL, with or without stromal coculture or FL. No bnx engraftment was seen 7–8 months after injection of cells transduced with viral supernatant and IL-3/IL-6/KL. The addition of Flk-2L resulted in engraftment of 3.7% of the cells, and stromal cocultured cells gave 5% engraftment, with $\sim 10\%$ G418 resistant hu-CFU-GM [82, 84, 122]. Inverse PCR analysis indicated that only 2–6 transduced HSC were responsible for lymphoid (T-cell) and myeloid engraftment [84]. Extension of the duration of observation to 12 months showed persistence of transduced cells and hu-CFU-GM [90]. Transfection with GALV-pseudotyped virus resulted in 27% of primary CFU-GM being G418-resistant, with 6% of human cells in the mouse BM at 12 months. These mice were injected with the equivalent of 5×10^5 CD34+ or 2×10^3 CD34+CD38− cells [85]. Comparison of G418-resistant CFU-GM in these mice at 12 months showed that, although 10% of CFU-GM were resistant following injection of CD34+ cells, only 0.6% of CFU-GM were G418-resistant following injection of CD38− cells.

These results indicate that long-term repopulating CD38– cells were not transduced. Addition of Flk-2L to the cytokine cocktail with stromal coculture increased the extent of bnx engraftment of transduced CD38– cells [85].

In the NOD/SCID model, quite high levels of human hematopoietic cell engraftment can be obtained by injection of 5×10^4 CB or 2×10^6 BM or mPB CD34+ cells into irradiated animals (300–400 rad). Under these conditions human CD45+ cells comprise 5–50% of the BM after 5–6 weeks. However, long-term studies are complicated by the very high incidence of lymphoma, which develops in these mice by 8 months of age. Larochelle et al. [120] reported only a low level of transduction of CB and BM-derived NOD/SCID repopulating cells, despite high levels of vector expression in primary CFU-GM (32–41%) and secondary 5 week CFU-GM (29% BM, 48% CB). By PCR analysis, only $\sim 2\%$ of total engrafted BM cells were transduced. In these studies, MNC were transduced by coculture with packaging cells, resulting in a progressive (up to 5-fold) decline in engraftment potential over 48 h, in part due to depletion of SRC by adherence to the producer cells. Recovery of the SRC was improved by a 48 h retronectin/viral supernatant protocol using CD34+ enriched populations that enabled 42% transduction of primary CFU. Nevertheless, only 1.4% of human CFU recovered from mouse BM at 6 weeks were transduced. Increasing the duration of culture with Retronectin increased the number of gene-marked cells, but beyond 3 days, repopulating capacity was rapidly lost. The interpretation of these studies was that (a) LTC-IC and SRC were different cell populations, with only the latter being 'true' stem cells, (b) SRC are refractory to transduction, possibly due to lack of viral receptor expression or failure to be rapidly activated into the cell cycle, and (c) as discussed above, conventional cytokine culture conditions do not maintain SRC in culture for a period sufficient to either activate them into cycle or effectively transduce them. More encouraging results have been reported in another study where a GALV/MLV vector was used to transduce CB CD34+CD38– cells for 72 h on MS-5 stroma, following a 72-hour priming with a combination of cytokines that included TPO and Flk-2L [46]. In that study, 60% of primary CFU-GM and 48% of 5-week secondary CFU-GM were transduced. Furthermore, 10% of NOD/SCID repopulating cells were transduced. Conneally et al. [124] primed CB CD34+ cells for 48 h with a cytokine cocktail that included KL and Flk-2L, and performed the transduction over full-length fibronectin with two cycles of virus supernatant over 48 h. A highly significant correlation was found between the transduction efficiency of LTC-IC (17%) and transduction of NOD/SCID repopulating cells. However, no correlation was found between transduction of primitive cells and transduction of primary CFU-GM transduction. Schilz et al. [116] used cytokine priming in the presence of anti-

TGF-β with a spinoculation and Retronectin transduction procedure to obtain very high transgene expression in cultured cells (97%). Sixty-three percent of primary CFU-GM, and 24% of secondary 5-week CFU-GM were NGFR+ upon transplantation into NOD/SCID mice. BM engraftment at 5–6 weeks ranged from 0.3 to 33.0% of human CD45+ cells, with 6–57% of these being NGFR+.

Conclusion

The early enthusiasm for retroviral gene therapy has been tempered by generally negative clinical results and a realization that many of the fundamental problems had not been overcome. Nevertheless, this mode of treatment becomes more feasible as each critical variable is identified and strategies are developed to overcome each problem. Unfortunately, no clinical studies addressing the majority of the potential variables have been performed. Developments in the field of experimental hematology, occurring concurrently with developments in vector design and improvements in transduction strategies, are now permitting HSC to be maintained or expanded in culture, with transduction efficiencies ($\sim 10\%$) that would be clinically significant for correction of genetic disease or for chemotherapy resistance. Debate still continues concerning efficient transduction of cells capable of long-term repopulation in humans. Recent studies point to loss of HSC and BM repopulating capacity following prolonged ex vivo cytokine exposure – the very conditions that optimized entry of quiescent stem cells into cycle. Certainly, the use of lentiviral vectors should reduce the need for prolonged in vitro priming and transduction (although activation from G_0 to G_1 may still be necessary in the case of HSC). An understanding of marrow homing mechanisms should also assist to overcome defects in this aspect of HSC function. The chemokine, stromal-derived factor-1 (SDF-1), produced by BM stromal cells, is a potent chemoattractant that promotes rapid transendothelial chemotaxis of primitive cells, including CD34+CD38−, LTC-IC, and CAFC [110]. Defects in chemotactic responsiveness of these primitive cells have been found following in vitro culture of CD34+ cells with cytokine combinations that include IL-3, which have been used extensively for priming cells for gene transfer. Defects in chemotaxis are particularly apparent with CD34+ cells of adult origin [110]. Until this problem is overcome, the duration of ex vivo manipulation of cells should be restricted, probably to no more than 3–5 days. In vivo evaluation in immunosuppressed mice and fetal sheep models, together with primate studies, are necessary, but the ultimate yardstick of success remains the well-controlled clinical trial.

References

1 Bodine DM, Moritz T, Donahue RE, Luskey BD, Kessler SW, Margin DI, Orkin SH, Nienhuis AW, Williams DA: Long-term in vivo expression of a murine adenosine deaminase gene in rhesus monkey hematopoietic cells of multiple lineages after retroviral mediated gene transfer into CD34+ bone marrow cells. Blood 1993;82:1975–1980.
2 Bordignon C, Notarangelo LD, Nobili N, Ferrari G, Casorati G, Panina P, Mazzolari E, Maggioni D, Rossi C, Servida P, Ugazio AG, Mavilio F: Gene therapy in peripheral blood lymphocytes and bone marrow for ADA-immunodeficient patients. Science 1995;270:470–475.
3 Kohn DB, Weinberg KJ, Nolta JA, Heiss LN, Lenarsky C, Crooks GM, Hanley ME, Annett G, Grooks JS, El-Khoureiy A, Lawrence K, Wells S, Moen RC, Nbastian J, Williams-Herman DE, Elder M, Wara D, Bowen T, Hershfied MS, Mullen CA, Blaese RM, Parkman R: Engraftment of gene-modified umbilical cord blood cells in neonates with adenosine deaminase deficiency. Nat Med 1995;1:1017–1023.
4 Raftopoulos H, Ward M, Bank A: High-level transfer and long-term expression of the human beta-globin gene in a mouse transplant model. Ann NY Acad Sci 1998;850:178–190.
5 Takekoshi KJ, Oh YH, Westerman KW, London IM, Leboulch P: Retroviral transfer of a human beta-globin/delta-globin hybrid gene linked to beta locus control region hypersensitive site 2 aimed at the gene therapy of sickle cell disease. Proc Natl Acad Sci USA 1995;92:3014–3018.
6 Sadelain M, Wang CH, Antoniou M, Grosveld F, Mulligan RC: Generation of a high-titer retroviral vector capable of expressing high levels of the human beta-globin gene. Proc Natl Acad Sci USA 1995;92:6728–6732.
7 Pawliuk R, Bachelot T, Raftopoulos H, Kalberer C, Humphries RK, Bank A, Leboulch P: Retroviral vectors aimed at the gene therapy of human beta-globin gene disorders. Ann NY Acad Sci 1998;850:151–162.
8 Xu L-C, Kluepfel-Stahl S, Blanco M, Schiffmann R, Dunbar C, Karlsson S: Growth factors and stromal support generate very efficient retroviral transduction of peripheral blood CD34+ cells from Gaucher patients. Blood 1995;86:141–146.
9 Migita M, Medin JA, Pawliuk R, Jacobson S, Nagle JW, Anderson S, Amiri M, Humphries RK, Karlsson S: Selection of transduced CD34+ progenitors and enzymatic correction of cells from Gaucher patients, with bicistronic vectors. Proc Natl Acad Sci USA 1995;92:12075–12079.
10 Donahue RE, Byrne ER, Kirby MR, Agricola BA, Sellers ES, Gaudernack G, Karlsson S, Lansdorp PM: Transplantation and gene transfer of the human glucocerebrosidase gene into immunoselected primate CD34+/Thy-1+ cells. Blood 1996;88:4166–4172.
11 Bauer TR, Schwartz BR, Liles C, Ochs HD, Hickstein DD: Retroviral-mediated gene transfer of the leukocyte integrin CD18 into peripheral blood CD34+ cells derived from a patient with leukocyte adhesion deficiency type I. Blood 1998;91:1520–1526.
12 Porter CD, Parkar MH, Collins MKL, Levinsky RJ, Kinnon C: Efficient retroviral transduction of human bone marrow progenitor and long-term culture-initiating cells: Partial reconstitution of cells from patients with X-linked chronic granulomatous disease by gp91-phox expression. Blood 1996;87:3722–3730.
13 Gilboa E, Smith C: Gene therapy for infectious disease: The AIDS model. Trends Genet 1994;10:139–144.
14 Mark CL, Yu M, Yamada O, Kraus G, Looney D, Poeschla E, Wong-Staal F: Transfer of an anti-HIV-1 ribozyme gene into primary human lymphocytes. Hum Gene Ther 1994;5:1115.
15 Sun LQ, Pyati J, Smythe J, Wang L, Macpherson J, Gerlach W, Symonds G: Resistance of human immunodeficiency virus type 1 infection conferred by transduction of human peripheral blood lymphocytes with ribozyme, antisense, or polymeric transactivation response element constructs. Proc Natl Acad Sci USA 1995;92:7272–7276.
16 Wang L, Witherington C, King A, Gerlach WL, Carr A, Penny R, Cooper D, Symonds G, Sun LQ: Preclinical characterization of an anti-tat ribozyme for therapeutic application. Hum Gene Ther 1998;9:1283–1291.
17 Sun LQ, Ely JA, Gerlach W, Symonds G: Anti-HIV ribozymes. Mol Biotech 1997;7:241–251.

18　Bahner I, Kearns K, Hao QL, Smorgorzewska E, Kohn DB: Transduction of human CD34+ hematopoietic progenitor cells by a retroviral vector expressing an RRE decoy inhibits human immunodeficiency virus type I replication in myelomonocytic cells produced in long-term culture. J Virol 1996;70:4352–4360.

19　Su L, Lee R, Bonyhadi M, Matsuzake H, Forestell S, Escaich S, Bohnlein E, Kaneshima H: Hematopoietic stem cell-based gene therapy for revM10 in myeloid cells in vivo and in vitro. Blood 1997;89:2283–2290.

20　Bauer G, Balder P, Kearns K, Bahner I, Wen SF, Zaia JA, Kohn DB: Inhibition of human immunodeficiency virus-1 (HIV-1) replication after transduction of granulocyte colony stimulating factor-mobilized CD34+ cells from HIV-1-infected donors using retroviral vectors containing anti-HIV-1 genes. Blood 1997;89:2259.

21　Jelinek J, Fairbairn LJ, Dexter TM, Rafferty JA, Stocking C, Ostertag W, Margison GP: Long-term protection of hematopoiesis against the cytotoxic effects of multiple doses of nitrosourea by retrovirus-mediated expression of human O6-alkylguanine-DNA-alkyltransferase. Blood 1996;87: 1957–1961.

22　Magni M, Shammah S, Schiro R, Mellando W, Dalla-Favera R, Gianni AM: Induction of cyclophosphamide-resistance by aldehyde-dehydrogenase gene transfer. Blood 1996;87:1097–1103.

23　Goldman MJ, Lee PS, Yang JS, Wilson JM: Lentiviral vectors for gene therapy of cystic fibrosis. Hum Gene Ther 1997;8:2261–2268.

24　Blomer U, Naldini L, Kafri T, Trono D, Verma IM, Gage FH: Highly efficient and sustained gene transfer in adult neurons with a lentivirus vector. J Virol 1997;71:6641–6649.

25　Naldini L, Blomer U, Gallay P, Ory D, Mulligan R, Gage FH, Verma IM, Trono D: In vivo gene delivery and stable transduction of nondividing cells by a lentiviral vector. Science 1996;272: 263–267.

26　Naldini L, Blomer U, Gage FH, Trono D, Verma IM: Efficient transfer, integration, and sustained long-term expression of the transgene in adult rat brains injected with a lentiviral vector. Proc Natl Acad Sci USA 1996;93:11382–11388.

27　Moore MAS, Hoskins I: Ex vivo expansion of cord blood derived stem cells and progenitors. Blood Cells 1994;20:468–481.

28　Piacibello W, Sanavio F, Garetto L, Severino A, Bergandi D, Farrario J, Fagoli F, Berger M, Aglietta M: Extensive amplification and cell-renewal of human primitive hematopoietic stem cells from cord blood. Blood 1997;89:2644–2653.

29　Wang JC, Lapidot T, Cashman JD, Doedens M, Addy L, Sutherland DR, Nayar R, Laraya P, Minden M, Keating A, Eaves AC, Eaves CJ, Dick JE: High level engraftment of NOD/SCID mice by primitive normal and leukemic hematopoietic cells from patients with chronic myeloid leukemia in chronic phase. Blood 1998;91:2406–2414.

30　Campain JA, Terrell KL, Tomczak JA, Shpall EJ, Hami LS, Harrison GS: Comparison of retroviral-mediated gene transfer into cultured human CD34+ hematopoietic progenitor cells derived from peripheral blood, bone marrow, and fetal umbilical cord blood. Biol Blood Marrow Transplant 1997;3:273–281.

31　Abe T, Ito M, Okamoto Y, Kim HJ, Takaue Y, Yasutomo K, Makimoto A, Yamaue T, Kawano Y, Watanabe T, Shimada T, Kuroda Y: Transduction of retrovirus-mediated NeoR gene into CD34+ cells purified from granulocyte colony-stimulating factor (G-CSF) mobilized infant and cord blood. Exp Hematol 1997;25:966–971.

32　Leonard JP, May C, Gallardo H, Raffi S, Sadelain M, Moore MAS: Retroviral transduction of human hematopoietic progenitor cells using a vector encoding a cell surface marker (LNGFR) to optimize transgene expression and characterize transduced cell populations. Blood 1996;88(suppl 1): 442a.

33　Cheng L, Du C, Lavau C, Chen S, Tong J, Chen BP, Scollay R, Hawley RG, Hill B: Sustained gene expression in retrovirally transduced, engrafting human hematopoietic stem cells and their lymphomyeloid progeny. Blood 1998;92:83–92.

34　Brenner MK, Rill DR, Holladay MS, Heslop HE, Moen RC, Buschle M, Krance RA, Santana VM, Anderson WF, Ihle JN: Gene marking to determine whether autologous marrow infusion restores long-term haematopoiesis in cancer patients. Lancet 1993;342:1134–1137.

35 Zanjani M, Amelia-Porada G, Livingstone AG, Flake AW, Ogawa M: Human bone marrow CD34– cells engraft in vivo and undergo multilineage expression that includes giving rise to CD34+ cells. Exp Hematol 1998;26:353–360.

36 Miller AD: Cell-surface receptors for retroviruses and implications for gene transfer. Proc Natl Acad Sci USA 1996;93:11407–11413.

37 Cone RD, Mulligan RC: High-efficiency gene transfer into mammalian cells: Genetration of helper-free recombinant retrovirus with broad mammalian host range. Proc Natl Acad Sci USA 1984;81: 6349–6363.

38 Danos O, Mulligan RC: Safe and efficient generation of recombinant retroviruses with amphotropic and ecotropic host ranges. Proc Natl Acad Sci USA 1988;85:6460–6464.

39 Miller AD, Garcia JV, von Suhr N, Lynch CM, Wilson C, Eiden MV: Construction and properties of retrovirus packaging cells based on gibbon ape leukemia virus. J Virol 1991;65:2220–2224.

40 Burns JC, Friedmann T, Driever W, Burrascano M, Yee J-K: Vesicular stomatitis virus G glycoprotein pseudotyped retroviral vectors: Concentration to very high titer and efficient gene transfer into mammalian and nonmammalian cells. Proc Natl Acad Sci USA 1993;89:8033–8037.

41 Orlic D, Girard LJ, Jordan CT, Anderson SM, Cline AP, Bodine DM: The level of mRNA encoding the amphotropic retrovirus receptor in mouse and human hematopoietic stem cells is low and correlates with the efficiency of retrovirus transduction. Proc Natl Acad Sci USA 1996;93:11097–11102.

42 Kiem H-P, Heyward S, Winkler A, Potter J, Allen JM, Miller AD, Andrews RG: Gene transfer into marrow repopulating cells: Comparison between amphotropic and gibbon ape leukemia virus pseudotyped retroviral vectors in a competitive repopulation assay in baboons. Blood 1997;90: 4638–4645.

43 Orlic D, Girard LJ, Anderson SM, Pyle LC, Yoder MC, Broxmeyer HE, Bodine DM: Identification of human and mouse hematopoietic stem cell populations expressing high levels of mRNA encoding retrovirus receptors. Blood 1998;91:3247–3254.

44 Crooks GM, Kohn DB: Growth factors increase amphotropic retrovirus binding to human CD34+ bone marrow progenitor cells. Blood 1993;82:3290–3297.

45 Kiem H-P, Andrews RG, Morris J, Peterson L, Heyward S, Allen JM, Rasko JEJ, Potter J, Miller AD: Improved gene transfer into baboon marrow repopulating cells using recombinant human fibronectin fragment CH-296 in combination with interleukin-6, stem cell factor, FLT-3 ligand, and megakaryocyte growth and development factor. Blood 1988;92:1878–1886.

46 Marandin A, Dubart A, Pflumbo F, Cosset F-L, Cordette V, Chapel-Fernandes S, Coulombel L, Vainchenker W, Louache F: Retrovirus-mediated gene transfer into human CD34+38 low primitive cells capable of reconstituting long-term cultures in vitro and nonobese diabetic-severe combined immunodeficiency mice in vivo. Hum Gene Ther 1998;9:1497–1511.

47 Takeuchi Y, Cosset F-LC, Lachman PJ, Okada H, Weiss RA, Collins MKL: Type C retrovirus inactivation by human complement is determined by both the viral genome and the producer cell. J Virol 1994;68:8001–8007.

48 Cosset F-L, Takeuchi Y, Battini J-L, Weiss RA, Collins MKL: High-titer packaging cells producing recombinant retrovirus resistant to human serum. J Virol 1995;69:7430–7436.

49 Russell DW, Berger MS, Miller AD: The effects of human serum and cerebrospinal fluid on retroviral vectors and packaging cell lines. Hum Gene Ther 1995;6:635–641.

50 Akkina RK, Walton RM, Chen ML, Li QX, Planelles V, Chen ISY: High efficiency gene transfer into CD34+ cells with a human immunodeficiency virus type 1-based retroviral vector pseudotyped with vesicular stomatitis virus envelope glycoprotein G. J Virol 1996;70:2581–2585.

51 Yam PY, Yee J-K, Ito JI, Sniecinski I, Doroshow JH, Forman SJ, Zaia JA: Comparison of amphotropic and pseudotyped VSV-G retroviral transduction in human CD34+ peripheral blood progenitor cells from adult donors with HIV-1 infection or cancer. Exp Hematol 1998;26:962–968.

52 Case SS, Price MA, Xu D, Bauer G, Naldini L, Verma IM, Kohn DB, Crooks GM: Stable transduction of CD34+/CD38– cells in extended long term culture using a lentiviral vector (abstract 413). Proc Am Soc Gene Ther 1998;104a.

53 Sutton RE, Wu HTM, Rigg R, Bohnlein E, Brown PO: Human immunodeficiency virus type 1 vectors efficiently transduce human hematopoietic stem cells. J Virol 1998;72:5781–5788.

54 Yee J-K, Miyanohara A, LaPort P, Bouic K, Burns JC, Friedmann T: A general method for the generation of high-titer, pantropic retroviral vectors: Highly efficient infection of primary hepatocytes. Proc Natl Acad Sci USA 1994;91:9564–9568.

55 Yang Y, Vanin EF, Whitt MA, Fornerod M, Zwart R, Schneiderman RD, Grosveld G, Nienhuis AW: Inducible, high-level production of infectious murine leukemia retroviral vector particles pseudotyped with vesicular stomatitis virus G envelope protein. Hum Gene Ther 1995;6:1203–1213.

56 Bertran J, Miller JL, Yang Y, Fenimore-Justman A, Rueda F, Vanin EF, Nienhuis AW: Recombinant adeno-associated virus-mediated high-efficiency, transient expression of the murine cationic amino acid transporter (ecotropic retroviral receptor) permits stable transduction of human HeLa cells by ecotropic retroviral vectors. J Virol 1996;70:6759–6766.

57 Scott-Taylor TH, Gallardo HF, Gansbacher B, Sadelain M: Adenovirus facilitated infection of human cells with ecotropic retrovirus. Gene Ther 1998;5:621–629.

58 MacKenzie KL, Hackett NR, Crystal RG, Moore MAS: Adenoviral vector-mediated gene transfer to primitive human hematopoietic progenitor cells; transduction and quantification of long-term culture-initiating cells, submitted.

59 Chinswangwatanakul W, Lewis JL, Manning M, Roberts IAG, Gordon MY: Use of G418 resistance to select cells retrovirally transduced with the NeoR gene. Exp Hematol 1998;26:185–186.

60 Movassagh M, Desmyter C, Baillou C, Chapel-Fernandes S, Guigon M, Klatzmann D, Lemoine FM: High-level gene transfer to cord blood progenitors using gibbon ape leukemia virus pseudotype retroviral vectors and an improved clinically applicable protocol. Hum Gene Ther 1998;9:225–234.

61 Humeau L, Chabannon C, Firpo MT, Mannoni P, Bagnis C, Roncarolo M-G, Namikawa R: Successful reconstruction of human hematopoiesis in the SCID-hu mouse by genetically modified, highly enriched progenitors isolated from fetal liver. Blood 1997;90:3496–3506.

62 Kiem HP, Darovsky B, von Kalle C, Goehle S, Graham T, Miller AD, Storb R, Scuening FG: Long-term persistence of canine hematopoietic cells genetically marked by retrovirus vectors. Hum Gene Ther 1996;7:89–93.

63 Conneally E, Bardy P, Eaves CJ, Thomas T, Chappel S, Shpall EJ, Humphries RK: Rapid and efficient selection of human hematopoietic cells expressing murine heat-stable antigen as an indicator of retroviral-mediated gene transfer. Blood 1996;87:456–464.

64 Valtieri M, Schiro R, Chelucci C, Masella B, Testa U, Casella I, Montesoro E, Mariani G, Hassah JH, Peschle C: Efficient transfer of selectable and membrane reporter genes in hematopoietic progenitor and stem cells purified from peripheral blood. Cancer Res 1994;54:4398–4404.

65 Ruggieri L, Aiuti A, Salomoni M, Zappone E, Ferrari G, Bordignon C: Cell-surface marking of CD34+-restricted phenotypes of human hematopoietic progenitor cells by retrovirus-mediated gene transfer. Hum Gene Ther 1997;8:1611–1623.

66 Hutchings M, Moriwaki K, Dilloo D, Hoffmann T, Kimbrough S, Johnsen HE, Brenner MK, Heslop HE: Increased transduction efficiency of primary hematopoietic cells by physical colocalization of retrovirus and target cells. J Hematother 1998;7:217–224.

67 McCowage GB, Phillips KL, Gentry TL, Hull S, Kurtzberg J, Gilboa E, Smith C: Multiparameter-fluorescence activated cell sorting analysis of retroviral vector gene transfer into primitive umbilical cord blood cells. Exp Hematol 1998;26:288–298.

68 Persons DA, Allay JA, Allay ER, Smeyne RJ, Ashmun RA, Sorrentino BP, Nienhuis AW: Retroviral-mediate transfer of the green fluorescent protein gene into murine hematopoietic cells facilitates scoring and selection of transduced progenitors in vitro and identification of genetically modified cells in vivo. Blood 1997;90:1777–1786.

69 Mazurier F, Moreau-Gaudry F, Maguer-Satta V, Salesse S, Pigeonnier-Lagarde V, Ged C, Belloc F, Lacombe F, Mahon FX, Reiffers J, de Verneuil H: Rapid analysis and efficient selection of human transduced primitive hematopoietic cells using the humanized S65T green fluorescent protein. Gene Ther 1998;5:556–562.

70 Limon A, Briones J, Puig T, Carmona M, Fornas O, Cancelas JA, Nadal M, Garcia J, Rueda F, Barquinero J: High-titer retroviral vectors containing the enhanced green fluorescent protein gene for efficient expression in hematopoietic cells. Blood 1997;90:3316–3321.

71 Grignani F, Kinsella T, Mencarelli A, Valtieri M, Riganelli D, Grignani F, Lanfrancone L, Peschle C, Nolan GP, Pelicci PG: High-efficiency gene transfer and selection of human hematopoietic progenitor cells with a hybrid EBV/retroviral vector expressing the gene fluorescence protein. Cancer Res 1998;58:14–19.

72 Miller DG, Adam MA, Miller AD: Gene transfer by retrovirus vectors occurs only in cells that are actively replicating at the time of infection. Mol Cell Biol 1990;10:4239–4242.

73 Lewis PF, Emerman M: Passage through mitosis is required for oncoretroviruses but not for the human immunodeficiency virus. J Virol 1994;68:510–516.

74 Emmons RVB, Doren S, Zujewski J, Cottler-Fox M, Carter CS, Hines K, O'Shaughnessy JA, Leitman SF, Greenblatt JJ, Cowan K, Dunbar CE: Retroviral gene transduction of adult peripheral blood or marrow-derived CD34+ cells for six hours without growth factors or on autologous stroma does not improve marking efficiency assessed in vivo. Blood 1997;89:4040–4046.

75 Flasshove M, Banerjee D, Leonard JP, Mineishi S, Li MX, Bertino JR, Moore MAS: Retroviral transduction of human CD34+ umbilical cord blood progenitor cells with a mutated dihydrofolate reductase cDNA. Hum Gene Ther 1998;9:63–71.

76 Flasshove M, Banerjee D, Mineishi S, Li ML, Bertino JR, Moore MAS: Ex vivo expansion and selection of human CD34+ peripheral blood progenitor cells after introduction of a mutated dihydrofolate reductase cDNA via retroviral gene transfer. Blood 1995;85:566–574.

77 Shapiro F, Yao TJ, Raptis G, Reich L, Norton L, Moore MA: Optimization of conditions for ex vivo expansion of CD34+ cells from patients with stage IV breast cancer. Blood 1994;84:3567–3574.

78 Shapiro F, Pytowski B, Raffi S, Witte L, Hicklin DJ, Yao TJ, Moore MAS: The effects of Flk-2/flt3 ligand as compared with c-kit ligand on short-term and long-term proliferation of CD34+ hematopoietic progenitors elicited from human fetal liver, umbilical cord blood, bone marrow, and mobilized peripheral blood. J Hematother 1996;5:655–662.

79 Petzer AL, Hogge DE, Lansdorp PM, Reid DS, Eaves CJ: Self-renewal of primitive human hematopoietic cells (long-term-culture-initiating cells) in vitro and their expansion in defined medium. Proc Natl Acad Sci USA 1996;93:1470–1474.

80 Petzer AL, Zandstra PW, Piret JM, Eaves CJ: Differential cytokine effects on primitive (CD34+CD38–) human hematopoietic cells: Novel responses to flt-3 ligand and thrombopoietin. J Exp Med 1996;183:2551–2558.

81 Zandstra PW, Conneally E, Petzer AL, Piret JM, Eaves CJ: Cytokine manipulation of primitive human hematopoietic cell self renewal. Proc Natl Acad Sci USA 1997;94:4698–4703.

82 Nolta JA, Smogorzewska EM, Kohn DB: Analysis of optimal conditions for retroviral-mediated transduction of primitive hematopoietic cells. Blood 1995;86:101–110.

83 Dunbar CE, Cottler-Fox M, O'Shaughnessy JA, Doren S, Carter C, Berenson R, Brown S, Moen RC, Greenblatt J, Stewart FM, Leitman SF, Wilson WH, Vowan K, Young NS, Nienhuis AW: Retrovirally marked CD34-enriched peripheral blood and bone marrow cells contribute to long-term engraftment after autologous transplantation. Blood 1995;85:3048–3057.

84 Dao MA, Shah AJ, Crooks GM, Nolta JA: Engraftment and retroviral markings of CD34+ and CD34+CD38– human hematopoietic progenitors assessed in immune-deficient mice. Blood 1998;91:1243–1255.

85 Dao MA, Hannum CH, Kohn DB, Nolta JA: Flt3 ligand preserves the ability of human CD34+ progenitors to sustain long-term hematopoiesis in immune-deficient mice afer ex vivo retroviral-mediated transduction. Blood 1997;89:446–456.

86 Hesforffer C, Ayello J, Ward M, Kaubisch A, Vahdat L, Balmaceda C, Garrett T, Fetell M, Reiss R, Bank A, Antman K: Phase I trial of retroviral-mediated transfer of the human MDR1 gene as marrow chemoprotection in patients undergoing high-dose chemotherapy and autologous stem-cell transplantation. J Clin Oncol 1998;16:165–172.

87 Veena P, Traycoff CM, Williams DA, McMahel J, Rice S, Cornetta K, Srour EF: Delayed targeting of cytokine-nonresponsive human bone marrow CD34+ cells with retrovirus-mediated gene transfer enhances transduction efficiency and long-term expression of transduced genes. Blood 1998;91:3693–3701.

88 Boezeman J, Raymakers R, Vierwinden G, Linssen P: Automatic analysis of growth onset, growth rate and colony size of individual bone marrow progenitors. Cytometry 1997;28:305–308.

89 Huang S, Francis K, Law P, Ramakrishna R, Palsson BO, Ho AD: Kinetics and symmetry of initial cell division of human CD34+/CD38– cells from different ontogenic age (abstract 45). Biol Blood Marrow Transplant 1998;4:107.
90 Dao MA, Yu XJ, Nolta JA: Clonal diversity of primitive human hematopoietic progenitors following retroviral marking and long-term engraftment in immune-deficient mice. Exp Hematol 1997;25:1357–1366.
91 Horwitz M, Malech HL, Anderson SM, Girard LJ, Bodine D, Orlic D: G-CSF mobilized peripheral blood CD34+CD38– hematopoietic stem cells (HSC) are candidate target cells for high efficiency retrovirus gene therapy (abstract 81). Proc Am Soc Gene Ther 1998;21a.
92 Dilber MS, Bjorkstrand B, Li KJ, Smith CIE, Xanthopoulos KG, Gahrton G: Basic fibroblast growth factor increases retroviral-mediated gene transfer into human hematopoietic peripheral blood progenitor cells. Exp Hematol 1994;22:1129–1133.
93 Elwood NJ, Zogos H, Willson T, Begley CG: Retroviral transduction of human progenitor cells: Use of granulocyte colony-stimulating factor plus stem cell factor to mobilize progenitor cells in vivo and stimulation by Flt3/Flk-2 ligand in vitro. Blood 1996;88:4452–4462.
94 Hennemann B, Conneally E, Leboulch P, Rose-John S, Kalberer CP, Humphries RK, Eaves CJ: Efficient gene transfer into human CD34+ cord blood cells and progenitors using a GALV pseudotyped recombinant retrovirus containing the neomycin resistance and the green fluorescent protein (GFP) genes (abstract 342). Proc Am Soc Gene Ther 1998;86a.
95 Tisdale JF, Hanazono Y, Sellers SE, Agricola BA, Metzger ME, Donahue RE, Dunbar CE: Ex vivo expansion of genetically marked rhesus peripheral blood progenitor cells results in diminished long-term repopulating ability. Blood 1998;92:1131–1141.
96 Mobest D, Junghann I, Fichtner I, Becker M, Just U, Mertelsmann R, Goan S, Henschler R: Kinetics of NOD/SCID mouse repopulating cells (SRC) diverge from colony-forming cells (CFC) and long-term culture-initiating cells (LTC-IC) during ex vivo expansion of CD34+ blood progenitor cells (PBC). Blood 1997;90(suppl 1):365abs.
97 Brugger W, Heimfeld S, Berenson RJ, Mertelsmann R, Kanz L: Reconstitution of hematopoiesis after high dose chemotherapy by autologous progenitor cells generated ex vivo. N Engl J Med 1995;333:283–287.
98 Holyoake TL, Alcorn MJ, Richmond L, Farrell E, Pearson C, Green R, Dunlop DJ, Fitzsimons E, Pragnell IB, Franklin IM: CD34+ PBPC expanded ex vivo may not provide durable engraftment following myeloablative chemoradiotherapy regimens. Bone Marrow Transplant 1997;19:1095.
99 Tavassoli M, Konno M, Shiota Y, Omoto E, Minguell JJ, Znajani ED: Enhancement of the grafting efficiency of transplanted marrow cells by preincubation with interleukin-3 and granulocyte-macrophage colony-stimulating factor. Blood 1991;77:1599–1606.
100 Van der Loo JCM, Ploemacher RE: Marrow- and spleen-seeding efficiencies of all murine hematopoietic stem cell subsets are decreased by preincubation with hematopoietic growth factors. Blood 1995;85:2598–2606.
101 Peters SO, Kittler ELW, Ramshaw HS, Quesenberry PJ: Ex vivo expansion of murine marrow cells with interleukin-3 (IL-3), IL-6, IL-11, and stem cell factor leads to impaired engraftment in irradiated hosts. Blood 1996;87:30–37.
102 Yonemura Y, Ku H, Hirayama F, Souza LM, Ogawa M: Interleukin 3 or interleukin 1 abrogates the reconstituting ability of hematopoietic stem cells. Proc Natl Acad Sci USA 1996;93:4040.
103 Matsunaga T, Hirayama F, Yonemura Y, Murray R, Ogawa M: Negative regulation by interleukin-3 (IL-3) of mouse early B-cell progenitors and stem cells in culture: Tranduction of negative signals by c and IL-3 proteins of IL-3 receptor and absence of negative regulation by granulocyte-macrophage colony-stimulating factor. Blood 1998;92:901.
104 Gan OI, Murdoch B, Larochelle A, Dick JE: Differential maintenance of primitive human SCID-repopulating cells, clonogenic progenitors, and long-term culture-initiating cells after incubation on human bone marrow stromal cells. Blood 1997;90:641–650.
105 Bhatia M, Bonnet D, Kapp U, Wang JC, Murdoch B, Dick JE: Quantitative analysis reveals expansion of human hematopoietic repopulating cells after short-term ex vivo culture. J Exp Med 1997;186:619.

106 Conneally E, Cashman J, Petzer A, Eaves C: Expansion in vitro of transplantable human cord blood stem cells demonstrated using a quantitative assay of their lympho-myeloid repopulating activity in nonobese diabetic-scid/scid mice. Proc Natl Acad Sci USA 1997;94:9836.

107 Spence SE, Keller JR, Ruscetti FW, McCauslin CS, Gooya JM, Funakoshi S, Longo DL, Murphy WJ: Engraftment of ex vivo expanded and cycling human cord blood hematopoietic progenitor cells in SCID mice. Exp Hematol 1998;26:507–514.

108 Shimizu Y, Ogawa M, Kobayashi M, Almeida-Porada G, Zanjani ED: Engraftment of cultured human hematopoietic cells in sheep. Blood 1998;91:3688.

109 Luens KM, Travis MA, Chen BP, Hill BL, Scollay R, Murray LJ: Thrombopoietin, kit ligand, and flk2/flt3 ligand together induce increased numbers of primitive hematopoietic progenitors from human CD34 + Thy-1 + Lin-cells with preserved ability to engraft SCID-hu bone. Blood 1998;91:1206.

110 Jo D-Y, Rafii S, Hamada T, Moore MAS: Transendothelial chemotaxis and marrow homing of hematopoietic stem cells are mediated by stromal cell-derived factor-1 (SDF-1) signaling through the chemokine receptor CXCR4. Blood, submitted.

111 Hatzfeld J, Li ML, Brown EL, Sookdeo H, Levesque JP, O'Toole T, Gurney C, Clark SC, Hatzfeld A: Release of early human hematopoietic progenitors from quiescence by antisense transforming growth factor beta 1 or Rb oligonucleotides. J Exp Med 1991;174:925–929.

112 Sitnicka E, Ruscetti FW, Priestley GV, Wolf NS, Bartelmez SH: Transforming growth factor β1 directly and reversibly inhibits the initial cell division of long-term repopulating hematopoietic stem cells. Blood 1996;88:82–88.

113 Heinrich MC, Zigler AJ, Chai L: TGF-beta regulated proteins bind to novel c-kit mRNA sequences. Blood 1996;88:541a.

114 Ohishi K, Katayama N, Itoh R, Mahmud N, Miwa H, Kita K, Minami N, Shirakawa S, Lyman SD, Shiku H: Accelerated cell-cycling of hematopoietic progenitors by the flt-3 ligand that is modulated by transforming growth factor-beta. Blood 1996;87:1718–1726.

115 Sansilvestri P, Cardoso AA, Batard P, Panterne B, Hatzfeld A, Lim B, Levesque JP, Monier MN, Hatzfeld J: Early CD34high cells can be separated into KIThigh cells in which transforming growth factor-beta (TGF-beta) downmodulates c-kit and KITlow cells in which anti-TGF-beta upmodulates c-kit. Blood 1995;86:1729–1735.

116 Schilz AJ, Brouns G, Knob H, Ottmann OG, Hoelzer D, Fauser AA, Thrasher AJ, Grez M: High efficiency gene transfer to human hematopoietic SCID-repopulating cells under serum-free conditions. Blood 1998;92:3163–3171.

117 Kearns K, Bahner I, Bauer G, Wei SF, Valdez P, Sheeler S, Woods L, Miller R, Casciato D, Galpin J, Church J, Kohn DB: Suitability of bone marrow from HIV-1 infected donors for retrovirus-mediated gene transfer. Hum Gene Ther 1997;8:301–311.

118 Hu R, Oyaizu N, Than S, Kalyanaraman VS, Wang XP, Pahwa S: HIV-1 gp160 induced transforming growth factor-β production in human PBMC. Clin Immunol Immunopathol 1996;80:283–286.

119 Chuck AC, Clarke MF, Palsson BO: Retroviral infection is limited by Brownian motion. Hum Gene Ther 1996;7:1527–1534.

120 Larochelle A, Vormoor J, Hanenberg H, Wang JCY, Bhatia M, Lapidot T, Moritz T, Murdoch B, Xiao XL, Kato L, Williams DA, Dick JE: Identification of primitive human hematopoietic cells capable of repopulating NOD/SCID mouse bone marrow: Implications for gene therapy. Nat Med 1996;2:1329–1331.

121 Dunbar CE, Seidel NE, Doren S, Sellers S, Cline AP, Metzger ME, Agricola BA, Donahue RE, Bodine DM: Improved retroviral gene transfer into murine and rhesus peripheral blood or bone marrow reproducing cells primed in vivo with stem cell factor and granulocyte colony-stimulating factor. Proc. Natl Acad Sci USA 1996;93:11871–11875.

122 Nolta JA, Dao MA, Wells S, Smogorzewska M, Kohn DB: Transduction of pluripotent human hematopoietic stem cells demonstrated by clonal analysis after engraftment in immune-deficient mice. Proc Natl Acad Sci USA 1996;93:2414–2418.

123 Andreadis S, Palsson BO: Kinetics of retrovirus mediated gene transfer: The importance of the intracellular half-life of retroviruses. J Theor Biol 1996;182:1–20.

124 Conneally E, Eaves CJ, Humphries RK: Efficient retroviral-mediated gene transfer to human cord blood stem cells with in vivo repopulating potential. Blood 1998;91:3487–3493.

125 Sekhar M, Kotani H, Doren S, Agarwal R, McGarrity G, Dunbar CE: Retroviral transduction of CD34-enriched hematopoietic progenitor cells under serum-free conditions. Human Gene Ther 1996;7:33–38.
126 Glimm H, Flugge K, Mobest D, Hofmann VM, Postmus J, Henschler R, Lange W, Finke J, Kiem HP, Schulz G, Rosenthal F, Mertelsmann R, von Kalle C: Hum Gene Ther 1998;10:771–778.
127 Kotani H, Newton PB III, Zhang S, Chiang YL, Otto E, Weaver L, Blaese RM, Anderson WF, McGarrity GJ: Improved methods of retroviral vector transduction and production for gene therapy. Hum Gene Ther 1994;5:19–28.
128 Bahnson AB, Dunigan JT, Baysal BE, Mohney T, Atchison RW, Nimgaonkar MT, Ball ED, Barranger JA: Centrifugal enhancement of retroviral mediated gene transfer. J Virol Methods 1995;54:131–143.
129 Emery DW, Wang XC, Andrews RG, Papayannopoulou T: Optimizing monkey models for bone marrow transduction with retrovirus vectors (abstract 344). Proc Am Soc Gene Ther 1998;87a.
130 Chuck AS, Palsson BO: Consistent and high rates of gene transfer can be obtained using flow-through transduction over a wide range of retroviral titers. Hum Gene Ther 1996;7:743–750.
131 Chuck AS, Palsson BO: Membrane adsorption characteristics determine the kinetics of flow-through transductions. Biotech Bioengineering 1996;51:260–270.
132 Eipers PG, Krauss JC, Palsson BO, Emerson SG, Todd RF III, Clarke MF: Retroviral-mediated gene transfer in human bone marrow cells grown in continuous perfusion culture vessels. Blood 1995;86:3754–3762.
133 Bertolini F, Battaglia M, Corsini C, Lazzari L, Soligo D, Zibera C, Thalmeier K: Engineered stromal layers and continuous flow culture enhance multidrug resistance gene transfer in hematopoietic progenitors. Cancer Res 1996;56:2566–2572.
134 Moritz T, Patel VP, Williams DA: Bone marrow extracellular matrix molecules improve gene transfer into human hematopoietic cells via retroviral vectors. J Clin Invest 1994;93:1451–1457.
135 Moritz T, Dutt P, Xiao X, Carstanjen D, Vik T, Hanenberg H, Williams DA: Fibronectin improves transduction of reconstituting hematopoietic stem cells by retroviral vectors: Evidence of direct viral binding to chymotryptic carboxy-terminal fragments. Blood 1996;88:855–862.
136 Hanenberg H, Xiao XL, Dilloo D, Hashino K, Kato I, Williams DA: Colocalization of retrovirus and target cells on specific fibronectin fragments increases genetic transduction of mammalian cells. Nat Med 1996;2:876–882.
137 Hanenberg H, Hashino K, Konishi H, Hock RA, Kato I, Williams DA: Optimization of fibronectin-assisted retroviral gene transfer into human CD34+ hematopoietic cells. Hum Gene Ther 1997;8:2193–2206.
138 Nolta JA, Hanley MB, Kohn DB: Sustained human hematopoiesis in immunodeficient mice by cotransplantation of marrow stroma expressing human interleukin-3: Analysis of gene transduction of long-lived progenitors. Blood 1994;83:3041–3051.

Malcolm A.S. Moore, DPhil, Memorial Sloan-Kettering Cancer Center,
1275 York Avenue (Mailbox 101), New York, NY 10021 (USA)
Tel. +1 212 639 7090, Fax +1 212 717 3618, E-Mail m-moore@ski.mskcc.org

Transfer of the MDR-1 Gene into Hematopoietic Cells

Arthur Bank, Maureen Ward, Charles Hesdorffer

Columbia University, New York, N.Y., USA

One approach to the cure of advanced cancer is by the use of more intensive chemotherapy. To avoid severe or life-threatening bone marrow toxicity accompanying this treatment, stem cell support in association with high-dose chemotherapy regimens is currently being used. However, even in this circumstance, the patient's stem cells and marrow capacity for regenerating normal hematopoiesis are often eventually severely limited with continued drug exposure. Our goal is to establish a chemotherapy-resistant population of bone marrow cells which can result in the use of higher doses of chemotherapy with less toxicity. Normal bone marrow cells express low levels of the human multiple drug resistance (MDR-1, MDR) gene, and thus, are susceptible to killing by classes of drugs that require the action of the MDR gene transmembrane protein, p-glycoprotein, for their export from cells; these drugs include the anthracyclines, vinca alkaloids, podophyllins and taxanes, all commonly used to treat cancer. We are trying to protect normal marrow cells from these drugs by retroviral gene transfer of the human MDR gene. If successful, exposure to these drugs should expand as well as protect this population of MDR-transduced cells.

There are limitations of this approach using chemoprotection of human hematopoietic stem cells (HSC) and progenitor cells (HPC) in the treatment of cancer. First, the dose-limiting toxicities of many MDR-responsive drugs are nonhematopoietic; for example, the cardiac toxicity of the anthracyclines and the neurologic toxicity of vincas limit their doses. Secondly, MDR drugs may not be true stem cell toxins and thus, HSC protection may not require increased MDR expression. This may be due, at least in part, to MDR p-glycoprotein expression in stem cells [1]. However, we have recently shown that up to 25% of MDR-expressing human HPC including BFU-E and CFU-

GM become resistant to doses of paclitaxel which kill over 95% of untransduced cells [2]. These findings strongly support a potential clinical role for successful MDR gene therapy by protecting HPC that express less p-glycoprotein, and, thus, reduce marrow complications, hospitalization and morbidity as red cells, granulocytes, and platelets are maintained. Other potential long-term goals of MDR gene therapy include the use of 2-gene vectors containing the MDR gene and another drug resistance gene, methyl guanine methyl transferase (MGMT), to increase the spectrum of drug resistance of marrow HSC to include the nitrosoureas. In addition, the marrow toxicity of drug combinations that include MDR and non-MDR-type drugs or two or more MDR-type drugs may be ameliorated by MDR gene therapy.

Bone marrow transplantation: Mononuclear cells harvested from peripheral blood by apheresis as well as marrow aspirates are capable of marrow reconstitution [3, 4]. These mononuclear cells have commonly been referred to as 'peripheral blood stem cells' (PBSC). We prefer their designation as peripheral blood progenitor cells (PBPC) since they contain large numbers of nonstem cell progenitors as well as stem cells. The use of PBPC provides several advantages over progenitors obtained from marrow: (1) they can be harvested by apheresis instead of marrow aspiration which requires general anesthesia; (2) repeated PBPC harvesting can be done; (3) marrow reconstitution with PBPC appears to be faster than that with marrow; (4) PBPC can be transduced at least as well as marrow-derived cells [2, 5, 6]. HPC is the term we will use to indicate populations of early- and large-stage nucleated hematopoietic cells, such as human CD34+ cells we believe contain most, if not all, of the HSC. HPC are defined here as cells containing the HSC. The criteria for defining a true HSC are still in dispute [7]. HSC are defined here as cells capable of self-renewal and long-term marrow reconstitution in vivo of all hematopoietic cell lineages (red cells, granulocytes, platelets, macrophages, and lymphocytes); HSC may also be required for short-term marrow repopulation.

It is now documented that the subset of CD34+ CD38– cells contains most of the human HSC, as assessed by long-term colony-initiating cell (LTC-IC) activity, and that these cells preferentially survive in NOD-SCID mice [8–10].

Retroviral gene transfer: The use of retroviral vectors is the method of choice for gene transfer into HPC and HSC since it is the only approach that consistently leads to high-level integration of transferred genes into host cell chromosomal DNA [11–14]. Retroviral integration ensures transfer of the added gene to all progeny when cells divide. This is in contrast to the use of adenoviruses which exist as episomes, and adeno-associated viruses vectors which inconsistently integrate into chromosomal DNA [13, 15, 16]. Retroviral helper cell lines, or packaging lines are available that produce *gag, pol* and *env* retroviral proteins, but have mutations that do not allow the encapsulation of an intact Moloney murine leukemia virus (Mo-MLV) genome [17–20]. The amphotropic packaging lines in these pseudotyped lines target the amphotropic receptor Ram-1 on murine cells and GLVR-2 (pit-2) on human and primate cells. By contrast, newer packaging lines use the gibbon ape leukemia virus (GALV) envelope to target GLVR-1 (GALVR, pit-1) receptors on target cells [21, 22]. Other packaging lines using the vesicular stomatitis virus (VSV) G protein (VSV-G) have been described that allow retroviral particles to be concentrated by centrifugation and high titer virus and increased retroviral transduction to be attained [23–25]. Tetracycline-inducible expression of VSV-G has been used to avoid the toxic effects of its overexpression in packaging lines [25].

Packaging lines transfected with retroviral vectors are called producer lines. The lack of generation of replication-competent retrovirus (RCR) in producer lines is required for safe gene transfer since it has been shown in primates that the presence of RCR results in insertional mutagenesis and lymphomas [26]. The packaging lines of GP+E86 (ecotropic) and GPAm-12 (amphotropic) developed in our laboratories have been extremely safe requiring at least 3 recombinations in a single cell to generate RCR [18, 19]. No evidence of RCR has been seen by us even in 15-liter batches of viral supernatants made for clinical use with these packaging lines [27].

Methods and Results

The human MDR gene has been most useful as a gene to assess the ability to transfer a selectable marker into human HPC and HSC and for potential human gene therapy. MDR p-glycoprotein is expressed at low levels in HPC and HSC. A highly active retroviral vector containing the human MDR cDNA in a Harvey retroviral backbone and driven by the retroviral viral LTR has been used in all of our studies [28, 29]. Specific oligonucleotide primers can be used to detect the MDR cDNA and its RNA product in cells by PCR analysis [29, 30]. Expression of the human MDR gene at the protein level is detected by antibodies [29, 31, 32].

We summarize our results using MDR gene transfer into murine and human cells and in a clinical trial below.

Use of Murine Marrow Cells to Optimize Human Gene Transfer into Human Progenitor Cells

We used our unique retroviral ecotropic and amphotropic packaging lines for safe and optimal gene transfer in these studies [18, 19]. These packaging cell lines contain the *gag-pol* and *env* genes on different plasmids to avoid recombination. The packaging cells lines that after transfection with an NeoR vector produced the highest titer ecotropic and amphotropic viruses are GP+E86 and GP+envAM12, respectively [18, 19]. These lines are both safe and highly efficient in gene transfer, and have been sent to >500 laboratories worldwide. Using these packaging lines, we have shown that the MDR gene can be successfully and stably transduced and expressed into mouse erythroleukemia cells, MELC [33] and the bone marrow cells of mice [29]. We first transfected the MDR cDNA-containing retroviral vector into our ecotropic and amphotropic packaging lines, and selected producer cells containing and expressing the MDR gene by exposure to colchicine. Cells that do not express the MDR gene are killed, while cells producing high levels of MDR survive. We isolated several surviving clones and used those with the highest viral titers: 5×10^5 viral particles/ml for the ecotropic line, and 5×10^4 viral particles/ml for the amphotropic MDR producer line. We utilized transfect/infect protocols to achieve our highest retroviral titers. In this scheme, we transfected the MDR-containing retroviral vectors into our GP+E86 ecotropic packaging cells, and selected for colchicine-resistant clones [29]. We then infected AM12 cells with these retroviral supernatants.

We cocultured our highest titer (5×10^5 particles/ml) ecotropic producer clone with mouse bone marrow cells, and transfused the transduced mouse bone marrow cells into the tail veins of lethally irradiated mice [29]. We demonstrated the presence of the human MDR gene in the peripheral blood of 90% of the transplanted mice using PCR analysis with human MDR-specific oligonucleotide primers 50 days after transplantation [29]. In other experiments, we examined the long-term expression of the human MDR gene in mice, and the cell types in which the inserted gene is expressed. At 8 months after transplantation, 50% of the successfully transplanted mice continued to contain the transduced human gene by PCR of peripheral blood [29]. Southern blots of bone marrow from sacrificed animals demonstrate the presence of the transferred gene.

To analyze the expression of the human MDR gene at the protein level, we used a monoclonal antibody, 17F9 [32], reacting with an external epitope of MDR p-glycoprotein. Using FACS analysis, we showed that clones of both producer cells and MELC transduced with the MDR retrovirus have a 1- to 2-log increase in MDR activity [29, 33]. In addition, a significant number of bone marrow cells from mice transduced with this retrovirus expressed high levels of MDR 8 months after transplantation. To rule out the possibility that long-lived lymphocytes were the source of the MDR-positive cells, cell gating on the basis of size and morphology was used to exclude lymphocytes from the analysis [29]. Approximately 14% of the granulocyte-macrophage cells in the bone marrow of a mouse 8 months after transplantation contain markedly increased amounts of MDR p-glycoprotein on their surface as compared to controls [29]. Since the lifespan of granulocytes is less than 2 weeks, these data showed that transduced cells in this granulocyte-macrophage population were derived from bone marrow stem cells transplanted 8 months earlier. As indicated previously, HPC express low levels of MDR p-glycoprotein and are preferentially killed by MDR-active drugs.

We showed that we can select in vivo for MDR-transduced cells in bone marrow [29]. Four transduced mice, initially positive by PCR for the MDR gene, subsequently lost their MDR PCR signal at 8 months after transplantation. These mice were given a single dose of paclitaxel in an attempt to enrich for bone marrow cells containing and expressing the human MDR gene, and to prove that MDR expression allows cell selection. Seven days after the dose of paclitaxel, the PCR signal in all four mice reappeared; in addition, FACS analysis of the peripheral blood of two of these mice showed 5–8% MDR-positive granulocytes indicating that we can select for MDR-expressing cells in vivo using drugs to enrich for these cells [29].

Thus, we demonstrated that: (1) Bone marrow stem cells can be stably transduced with the human MDR gene and remain active for long periods of time, and (2) MDR-transduced cells are protected from paclitaxel toxicity and can be selected.

Use of Mouse Fetal Liver Cells to Define the Optimal Conditions for Transfer of Human Genes into Human Progenitor Cells

We have used the mouse fetal liver cells (FLC) as a convenient source of HPC and HSC to study methods of enhancing marrow reconstitution with transduced cells containing and expressing a transferred human gene, again the MDR gene, in the ablated mouse model [29, 34, 35]. In these studies, we first found that FLC lack amphotropic receptor expression [34]. We then showed very efficient ecotropic MDR retroviral gene transfer and expression of transduced FLC in mice for up to 1 year after transplantation, indicating stem cell transduction, and long-term maintenance and p-glycoprotein expression of the transduced cells [34, 35]. We also studied the fate of different cell types in FLC in marrow reconstitution in mice using MDR gene marking of hematopoietic subpopulations isolated from FLC [35]. Subpopulations were isolated using the antibodies, anti-Thy-1 and Sca-1, as well as a panel of lineage-specific antibodies to isolate Lin– cells. In these studies, we found that populations enriched for HSC (Lin–, Thy-1^{low} and Sca-1 +cells), also had an increased capacity to repopulate marrow-ablated mice with marked cells long-term. These results suggested that similar enrichments for subpopulations of human marrow cells may be useful to improve the long-term repopulating activity (LTRA) of human HPC [35]. These studies demonstrated that highly enriched HPC are efficient at long-term repopulation of mouse marrow.

In other experiments, preselection of MDR-expressing FLC by FACS resulted in much higher percentages of p-glycoprotein-expressing marrow cells, especially long-term as compared to reinfusion of the unfractionated FLC population [35]. These results demonstrate that isolation of HPC-enriched cells and of p-glycoprotein expression cells prior to marrow reconstitution are potentially useful in increasing the efficiency of transfer and expression of foreign genes. However, it should be noted that only oligoclonal reconstitution was achieved, and that the murine ecotropic receptor was targeted by virus in these experiments.

We have also analyzed the expression of the murine amphotropic receptor, Ram-1, on hematopoietic cells during mouse development [34, 36]. In these studies, we demonstrated that Ram-1 is not expressed on yolk sac cells at days 9.5–11.5 postcoitum (p.c.); Ram-1 begins to be expressed in FLC between days 13.5 and 14.5 p.c., and is highly expressed in adult mouse bone marrow and spleen [36]. We also showed in these experiments that the expression of Ram-1 on stem-cell-enriched populations of Sca+ Lin− Thy^{low} cells was markedly reduced as compared to later cells, an indication that amphotropic receptor on HSC might be much lower than ecotropic receptor on murine cells.

Upregulation of the Amphotropic Retroviral Receptor on Mouse Cells and Human CD34+ Cells

Since Ram-1 is known to be a phosphate transporter, we also studied its possible upregulation with phosphate deprivation [21]. We found that in cell culture, phosphate deprivation leads to increased Ram-1 expression in 14.5 day FLC; in addition, phosphate deprivation leads to the ability of these cells to be transduced by amphotropic MDR retrovirus in contrast to their inability to do so in normal phosphate medium [36]. Since upregulation of the human amphotropic receptor, GLVR-2 on human HPC and HSC might be an additional means of increasing MDR gene transduction into these cells, we have been measuring the expression of the amphotropic receptor on human CD34+ cells under different conditions including phosphate deprivation [37, 38]. In these studies, CD34+ cells have been isolated by CellPro columns, and placed in culture with IL-3, IL-6 and SCF for varying times up to 3 days in normal or low-phosphate-containing media. Over 3 days of culture, there is a 2- to 3-fold increase in GLVR-2 expression, as measured by competitive RT-PCR using actin as a control [37, 38]. There is an additional 2-fold increase in GLVR-2 expression using this assay when the cells are incubated in low-phosphate medium [37, 38]. These studies are being pursued as an approach to increasing transduction efficiency of human HSC.

Human Marrow Transduction with a Human Gene Using a Safe MDR Amphotropic Producer Line

We have also transduced human bone marrow cells with supernatants from our highest titer amphotropic MDR producer line, A12M1, and analyzed the results by PCR, and FACS for MDR and by in vitro analysis of methylcellulose colonies for progenitors [2, 6]. Supernatants from A12M1 have been used instead of coculture with MDR producer cells to avoid potential contamination of the HPC with the producer cells, an undesirable event in clinical use. In these studies, we have shown that both Ficoll-separated mononuclear cells, and CD34+ cells can be efficiently transduced after a 48-hour preincubation with growth factors IL-3, IL-6 and SCF, and two changes of MDR retroviral supernatant over the subsequent 24 h at a ratio of viral particles to cells of greater than 1 [2, 6]. We have used fibronectin plates to culture human CD34+ cells. CD34+ cells were isolated from bone marrow using either: (1) Ficoll gradient centrifugation to obtain mononuclear cells followed by negative cell selection on soybean agglutinin-coated plates (Applied Immune Sciences, AIS) and then CD34+ cell selection using anti-CD34-antibody-coated plates (AIS); or (2) CellPro Ceprate columns [39, 40]. Both techniques yield high percentages of functional CD34+ cells from bone marrow as assessed for progenitor enrichment by methylcellulose assays for BFU-E and CFU-GM; however, the CellPro columns are easier to use. Fifty- to 100-fold enrichment for these progenitors is routinely obtained. Up to 60% of BFU-E and CFU-GM derived from these transduced cells contain the transferred MDR gene, and up to 11% of the CD34+ cells expanded by exposure to G-CSF and GM-CSF express increased amounts of MDR by FACS analysis using an anti-MDR antibody. Paclitaxel exposure selects for MDR-transduced cells [6].

We have used the LTC-IC assay [41] to compare LTRA of retrovirally transduced and untransduced marrow cells, and found that transduction does not decrease the number of LTC-IC [6]. In addition, 25% (4 of 16) of individual CFU-GM derived from LTC-IC were transduced, as demonstrated by PCR analysis in this study.

We have also used our A12M1 producer lines to show high levels of MDR transduction into cord blood cells and using stromal layers [42].

MDR Transduction of Human Peripheral Blood Progenitor Cells

CD34+ cells obtained from PBPC can be transduced with amphotropic MDR retroviral supernatants at least as well as marrow-derived CD34+ cells [2]. In these experiments, we showed that preincubation of CD34+ cells in the presence of growth factors IL-3, IL-6 and SCF is required for optimal MDR transduction, as assessed by the number of transduced BFU-E and CFU-GM [2]. Eliminating either the time of preincubation or the presence of growth factors results in significantly less PBPC CD34+ cell transduction [2]. Most importantly, we showed that in the presence of both preincubation and growth factors, up to 25% of BFU-E and CFU-GM obtained from transduced CD34+ cells are protected from a dose of paclitaxel (10^{-8} M) which kills over 95% of untransduced cells [2]. These studies demonstrate that protection of human HPC by MDR gene transfer from the toxic effects of paclitaxel is feasible.

The Use of the Human MRP Gene as a Selectable Chemoprotectant

The multidrug resistance (MRP) gene is another potential gene of interest in protocols whose goal is to chemoprotect marrow at the same time that reversal of MDR drug expression in tumors is attempted. The MRP gene encodes a protein with a similar spectrum of drug resistance to that of MDR [43, 44]. The MRP cDNA is cloned, and is only expressed at low levels in normal marrow cells, and may, like MDR, induce resistance to drug toxicity when transferred and expressed in marrow cells. Unlike MDR p-glycoprotein, the MRP efflux pump activity is not inhibited by MDR reversal drugs like verapamil and cyclosporine which inhibit the action of MDR. The MRP gene may thus be uniquely useful together with MDR-reversal agents. The protective effect of MDR marrow gene therapy would be reversed with MDR reversal drugs since they would decrease the gene transfer-induced MDR levels in marrow as well as in the tumors. By contrast, MRP gene therapy into marrow cells could be used simultaneously with the MDR reversal drugs. This strategy might make it possible to lower MDR expression in tumors, while MRP action in marrow cells is maintained and prevents MDR drug-induced myelotoxicity. We have shown that 3T3 cells containing an MRP cDNA-containing retroviral vector have increased resistance to Adriamycin and vincristine [45]. More recently, MRP gene transfer has been demonstrated in mouse hematopoietic cells [46].

A Phase-1 Human Gene Therapy MDR Clinical Trial

A phase-1 clinical trial has begun based on the efficacy and safety of MDR gene transfer we demonstrated in the preclinical studies described above. One goal of this protocol is to use MDR gene transfer to protect human HSC and HPC from the myelotoxicity of MDR-active drugs. This could lead to the subsequent administration of more of these drugs with less marrow side effects. This protocol was approved by the Recombinant DNA Committee and the FDA using MDR retroviral supernatants from the safe amphotropic MDR producer line, A12M1 [47]. In this first trial, up to 20 patients with advanced breast, ovarian and brain cancer without marrow involvement are receiving MDR-transduced CD34+ cells as part of a protocol in which they are also undergoing ABMT and intensive chemotherapy for their tumors. The A12M1 cells and supernatants are extensively tested and shown to be free of all known pathogens, and to have no RCR. Up to one-third of the marrow is used for CD34+ cell isolation, and is MDR-transduced, while the rest of the marrow or PBPC reinfused are unmanipulated and reinfused with the transduced cells. The end points of the phase-1 protocol are: (1) clinical side effects and the presence of RCR; (2) efficacy and stability of marrow MDR transduction, and (3) enrichment of the marrow for transduced cells in those patients subsequently receiving paclitaxel.

We have recently published the results to date of this first clinical trial in 7 patients using the growth factor combination IL-3, IL-6 and SCF [27]. Two patients with breast cancer progressed after registration and marrow harvesting and were removed from the study; one because of the presence of metastatic disease noted in the marrow harvest, and the other for bone metastases. These 2 patients underwent high-dose therapy but received only untransduced cells for hematopoietic support. The other 5 patients underwent high-dose therapy with reinfusion of the transduced cells. Toxicities were those expected for dose-intensive therapy requiring stem cell support (pancytopenia, fever, diarrhea, mucositis, multiorgan failure). No toxicities attributable to the use of transduced stem cells were observed. Length of hospital stay, time to engraftment and blood product requirements after transplantation were comparable to those of patients not treated with MDR-transduced cells [27]. Although time to engraftment for one patient was prolonged, he had received carmustine (BCNU), a stem cell poison, for his brain tumor prior to marrow harvesting. It is notable that another patient initially in the study, who was also heavily pretreated with drugs including cisplatin prior to undergoing her course of high-dose chemotherapy, had a protracted period of pancytopenia although she did not receive any transduced cells.

Marrow was used for transduction for the first 6 patients entered on this study, although both marrow and PBPC were reinfused to ensure prompt engraftment. After the protocol was amended, patient 7 had only PBPC harvested, a portion of these being transduced and the remainder cryopreserved unmanipulated. Notably, PCR of marrow of patients after transduction and before transplantation showed that even with the scale-up of retroviral supernatant and CD34+ cells used, 20–70% of BFU-E and CFU-GM derived from transduced CD34+ cells were positive by MDR PCR [27]. This indicates the feasibility of scaling up the transduction protocol in clinical trials without losing the transduction efficiency of HPC.

However, PCR analysis of patient marrow samples after transplantation for the transferred exogenous MDR cDNA was positive in one sample each from 2 patients 3 weeks and 4 months after transplantation, respectively [27].

Semiquantitation of the MDR PCR signal with control samples of A12M1 diluted to varying extents revealed that less than 1:1,000 cells in the samples were transduced. PCR positivity was observed in the 2 patients reinfused with the highest percentage of MDR-transduced cells, 25 and 26%, respectively, as compared to lower percentages in the other patients. Unfortunately, both MDR-positive patients died early and could not be followed for persistence of their transduced cells. FACS analysis of patient samples for increased p-glycoprotein expression, and methylcellulose analysis of individual BFU-E and CFU-GM by MDR PCR have all been negative. Two breast cancer patients given transduced cells have not exhibited any disease progression, making further chemotherapy unnecessary at present, and are continuing to be monitored for MDR gene transfer and expression.

One percent of the transduced cells and 5% of the suspension media was collected and evaluated by the National Gene Vector Laboratory (NGVL) in Indianapolis for the presence of RCR. This material, as well as samples of blood and marrow harvested from the patients at 2 weeks, 4 weeks, and 3, 6 and 9 months after transplantation (when available), was assayed by coculture with *Mus dunni* cells for 4 weeks, followed by PG4 S+L− assay, and all samples have thus far been negative for RCR in this sensitive assay [27].

In summary, in this phase-1 trial, to date we have demonstrated the safety of the transduction process itself with respect to normal marrow engraftment, and have shown that we can attain high-level MDR transduction of colonies derived from CD34+ cells after large-scale clinically applicable protocol preparation. In order to improve HSC MDR transduction, we intend to define and utilize a more optimal growth factor transduction cocktail containing cytokines such as Flt3 and thrombopoietin (TPO, MGDF). We conclude that gene transduction of hematopoietic progenitor cells by retroviral vectors is feasible, that hematopoietic engraftment occurs successfully, and that the gene transfer can be detected after marrow recovery. No untoward toxicities related to gene transduction are apparent. We believe we have laid the foundation for further clinical studies in which the optimal clinical efficacy of MDR gene therapy can be evaluated; however, we have not yet achieved clinically meaningful success.

The Major Problem: The Human Hematopoietic Stem Cell

We believe that high-level human MDR gene transfer and expression in human HSC is a requirement for successful MDR bone marrow gene therapy. Our preclinical and clinical results suggest that HSC transduction is required for both short-term as well as long-term marrow repopulation in human and murine marrow reconstitution. We have had success in defining the conditions for transducing murine HSC with the human MDR gene using exogenous cytokines [29, 35]. However, in these experiments which show that some murine HSC have LTRA, we have used an ablated mouse model, and oligoclonal repopulation by only a few transduced stem cells is responsible for this positive result. In addition, there is no competition for repopulation by host HSC in this model. We were quite successful in transducing human CD34+ cells obtained from bone marrow and PBPC with the human MDR gene as assessed by methylcellulose assays for BFU-E and CFU-GM as well as LTC-IC [2, 6]. However, we now believe that these assays do not reflect human HSC transduction efficiency. We also conclude that these later transduced cells cannot repopulate the marrow efficiently, if at all. Despite a high proportion of transduced progenitors derived from CD34+ cells, even after scale-up for the clinical trial, we could only confirm the work of others that cytokine-treated cells engraft poorly [12, 27, 48, 49]. This is due to: (1) low-level inadequate transduction of human HSC because of their lack of adequate amphotropic receptors on these cells as compared to later progenitors; (2) a change in the biology of the human HSC after exposure to the exogenous cytokines used; and/or (3) the unfavorable competition for marrow repopulation of the cytokine-treated cells when infused with unmanipulated untransduced cells (to insure optimal marrow reconstitution). As mentioned earlier, growth factor-stimulated cells, at least after exposure to IL-3, IL-6, SCF and IL-11, are reported to have a defect in their LTRA when competed with unmanipulated donor cells in both marrow-ablated mice, and in unablated mice [50, 51]. Thus, we must overcome these obstacles to human HSC transduction and the effects of cytokine exposure to be successful in HSC gene therapy.

We plan to improve the conditions for the transduction of human HSC by upregulating the human amphotropic receptor, GLVR-2 or using other envelope-like genes such as VSV-G. We also hope to find and use growth factors and transduction conditions which permit HSC to divide and still retain their stem cell properties and remain capable of marrow reconstitution. Our A12M1 producer line has been used to prepare safe, FDA-approved clinical-grade supernatants, and, thus, any success we have establishing appropriate ex vivo culture conditions with cytokines or upregulation of GLVR-2 for MDR retroviral transduction with A12M1 supernatants can be rapidly applied in phase-1 clinical trials. If this is unsuccessful, the use of lentiviral components in new packaging and producer lines may be required since their use does not require cell division for retroviral integration as do the MLV-based viruses we utilized [52]. Since the majority of HSC are quiescent, this may be a critical advance. Obviously, safety issues must be addressed before any lentiviral lines can be used in clinical trials.

Conclusion

In summary, we have demonstrated successful human MDR gene transfer into mouse hematopoietic cells including stem cells with long-term repopulation of these cells in mice and the ability to select and expand the transduced cells with drugs. We have also demonstrated the ability for high-level MDR gene transfer and expression in human hematopoietic progenitors obtained from peripheral blood and bone marrow. In a phase-1 clinical trial, we have demonstrated the safety and feasibility of human MDR gene transfer in patients. However, the level of MDR gene transfer into hematopoietic cells in vivo is quite low, suggesting either a lack of gene transfer into HSC and/or a change in the biology of transduced stem cells which does not permit their efficient marrow repopulation. We assume that these stem cells are required for both short- and long-term marrow reconstitution. We know we can transduce and expand later human progenitors, but we must await appropriate conditions for transduction and repopulation of stem cells before we can attain a transduced progenitor and mature blood cell population derived from stem cells in patients which allows their chemoprotection.

References

1. Chaudhary PM, Roninson IB: Expression and activity of P-glycoprotein, a multidrug efflux pump, in human hematopoietic stem cells. Cell 1991;66:85–94.
2. Ward M, Pioli P, Ayello J, Reiss R, Urzi G, Richardson C, Hesdorffer C, Bank A: Retroviral transfer and expression of the human multiple drug resistance (MDR) gene in peripheral blood progenitor cells. Clin Cancer Res 1996;2:873–876.
3. Kessinger A, Armitage JO: The evolving role of autologous peripheral stem cell transplantation following high-dose therapy for malignancies (editorial). Blood 1991;77:211–213.
4. Pierelli L, Iacone A, Quaglietta AM, Nicolucci A, Menichella G, Benedetti Panici P, D'Antonio D, De Laurenzi A, De Rosa L, Fioritoni G, et al: Haematopoietic reconstitution after autologous blood stem cell transplantation in patients with malignancies: A multicentre retrospective study. Br J Haematol 1994;86:70–75.
5. Bregni M, Magni M, Siena S, Di-Nicola M, Bonadonna G, Gianni AM: Human peripheral blood hematopoietic progenitors are optimal targets of retroviral-mediated gene transfer. Blood 1992;80:1418–1422.
6. Ward M, Richardson C, Pioli P, Smith L, Podda S, Goff S, Hesdorffer C, Bank A: Transfer and expression of the human multiple drug resistance gene in human CD34+ cells. Blood 1994;84:1408–1414.
7. Orlic D, Bodine DM: Editorial: What defines a pluripotent hematopoietic stem cell (PHSC): Will the real PHSC please stand up! Blood 1994;84:3991–3994.
8. Terstapen LWMM, Huang M, Stafford M, et al: Sequential generations of hematopoietic colonies derived from single nonlineage committed CD34+ CD38– progenitor cells. Blood 1991;77:1218–1227.
9. Vormoor J, Lapidot T, Pflumio F, Risdon G, Patterson B, Broxmeyer H, Dick JH: Immature human cord blood progenitors engraft and proliferate to high levels in severe combined immunodeficient mice. Blood 1994;83:2489–2497.

10 Wang JCY, Doedens M, Dick JE: Primitive human hematopoietic cells are enriched in cord blood compared with adult bone marrow or mobilized peripheral blood as measured by the quantitative in vivo SCID-repopulating cell assay. Blood 1997;89:3919–3924.

11 Bank A, Ward M, Richardson C, Podda S, Smith L, Hesdorffer C: Retroviral gene transfer into hematopoietic stem cells: The human MDR gene as a model system; in Levitt D, Mertlesmann R (eds): Hematopoietic Stem Cells. New York, Marcel Dekker, 1995, pp 229–243.

12 Dunbar CE: Gene transfer to hematopoietic stem cells: Implications for gene therapy of human disease. Annu Rev Med 1996;47:11–20.

13 Brenner MK: Gene transfer to hematopoietic cells. N Engl J Med 1996;335:337–340.

14 Bank A: Human somatic cell gene therapy. Bioessays 1996;18:999–1007.

15 Crystal RG: Transfer of genes to humans: Early lessons and obstacles to success. Science 1995;270: 404–410.

16 Wilson JM: Adenoviruses as gene-delivery vehicles. N Engl J Med 1996;334:1185–1189.

17 Cone R, Mulligan RC: High-efficiency gene transfer into mammalian cells: Generation of helper-free recombinant retrovirus with broad mammalian host range. Proc Natl Acad Sci USA 1984;81: 6349–6353.

18 Markowitz D, Goff S, Bank A: A safe packaging line for gene transfer: Separating viral genes on two different plasmids. J Virol 1988;62:1120–1125.

19 Markowitz D, Goff S, Bank A: Construction and use of a safe and efficient amphotropic packaging cell line. Virology 1988;167:400–405.

20 Miller A: Progress toward human gene therapy. Blood 1990;76:271–278.

21 Kavanaugh MP, Miller DG, Zhang W, Law W, Kozak SL, Kabat D, Miller AD: Cell-surface receptors for gibbon ape leukemia virus and amphotropic murine retrovirus are inducible sodium-dependent phosphate symporters. Proc Natl Acad Sci USA 1994;91:7071–7075.

22 Kiem H-P, Heyward S, Winkler A, Potter J, Allen JM, Miller AD, Andrews RG: Gene transfer into marrow repopulating cells: Comparison between amphotropic and gibbon ape leukemia virus pseudotyped retroviral vectors in a competitive repopulation assay in baboons. Blood 1997;90: 4638–4645.

23 Yee JK, Miyanohara AA, Laporte P, Bouic K, Burns JC, Friedmann T: A general method for the generation of high-titer pantropic retroviral vectors: Highly efficient infection of primary hepatocytes. Proc Natl Acad Sci USA 1994;91:9564–9568.

24 Yang Y, Vanin E, Whitt M, Fornerod M, Zwart R, Schneiderman R, Grosveld G, Nienhuis A: Inducible, high-level production of infectious murine leukemia retroviral vector particles pseudotyped with vesicular stomatitis virus G envelope protein. Hum Gene Ther 1995;6:1203–1213.

25 Ory DS, Neugeboren BA, Mulligan R: A stable human-derived packaging cell line for production of high titer retrovirus/vesicular stomatitis virus G pseudotypes. Proc Natl Acad Sci USA 1996;93: 11400–11406.

26 Donahue RE, Kessler SW, Bodine D, McDonagh K, Dunbar C, Goodman S, Agricola B, Byrne E, Raffeld M, Moen R, et al: Helper virus induced T cell lymphoma in nonhuman primates after retroviral mediated gene transfer. J Exp Med 1992;176:1125–1135.

27 Hesdorffer C, Ayello J, Ward M, Reiss R, Vahdat L, Fetell M, Garrett T, Bank A, Antman KA: A Phase 1 trial of retroviral-mediated transfer of the human MDR1 gene as marrow chemoprotection in patients undergoing high-dose chemotherapy and autologous stem cell transplantation. J Clin Oncol 1998;16:165–178.

28 Pastan I, Gottesman MM: Multidrug resistance. Annu Rev Med 1991;42:277–286.

29 Podda S, Ward M, Himelstein A, Richardson C, de la Flor-Weiss E, Smith L, Gottesman M, Pastan I, Bank A: Transfer and expression of the human multiple drug resistance gene into live mice. Proc Natl Acad Sci USA 1992;89:9676–9680.

30 Noonan KE, Beck D, Holzmayer TA, Chin JE, Wunder JS, Andrulis IL, Gazdar AF, Willman CL, Griffith B, Von Hoff DD, Roninson I: Quantitative analysis of MDR1 (multidrug resistance) gene expression in human tumors by polymerase chain reaction. Proc Natl Acad Sci USA 1990; 87:7160–7164.

31 Kartner N, Evernden-Porelle D, Bradley G, Ling V: Detection of P-glycoprotein in multidrug-resistant cell lines by monoclonal antibodies. Nature 1985;316:820–823.

32   Aihara M, Aihara Y, Schmidt-Wolf G, Schmidt-Wolf I, Sikic BI, Blume KG, Chao NJ: A combined approach for purging multidrug-resistant leukemic cell lines in bone marrow using a monoclonal antibody and chemotherapy. Blood 1991;77:2079–2084.
33   de la Flor-Weiss E, Richardson C, Ward M, Himelstein A, Smith L, Podda S, Gottesman M, Pastan I, Bank A: Transfer and expression of the human multiple drug resistance gene into live mice. Blood 1992;80:3106–3111.
34   Richardson C, Ward M, Podda S, Bank A: Mouse fetal liver cells lack functional amphotropic retroviral receptors. Blood 1994;84:433–439.
35   Richardson C, Bank A: Preselection of transduced murine hematopoietic stem cell populations leads to increased long-term stability and expression of the human multiple drug resistance gene. Blood 1995;86:2579–2589.
36   Richardson CA, Bank A: Developmental-stage-specific expression and regulation of an amphotropic retroviral receptor in hematopoietic cells. Mol Cell Biol 1996;18:4240–4247.
37   Kaubisch A, Ward M, Hesdorffer C, Bank A: Up-regulation of the human amphotropic receptor, GLVR-2 in human CD34+ cells (abstract). Blood 1997;90(suppl):117a.
38   Kaubisch A, Ward M, Hesdorffer C, Bank A: Up-regulation of the amphotropic retroviral receptor GLVR-2 on human CD34+ cells, submitted.
39   Lebkowski JS, McNally MA, Finch S, Manzagol S, Pletcher A, Schain LR, Okarma TB: Enrichment of murine hematopoietic stem cells. Reconstitution of syngeneic and haplotype-mismatched mice. Transplantation 1990;50:1019–1027.
40   Berenson RJ, Bensinger WI, Hill R, Andrews RG, Garcia-Lopez J, Kalamasz DF, Still BJ, Buckner CD, Bernstein ID, Thomas ED: Stem cell selection – Clinical experience. Prog Clin Biol Res 1990; 333:403–410.
41   Sutherland HJ, Eaves CJ, Dragowska W, Lansdorp PM: Characterization and partial purification of human marrow cells capable of initiating long-term hematopoiesis in vitro. Blood 1989;74: 1563–1570.
42   Bertolini F, Monte LD, Corsini C, Lazzari L, Lauri E, Soligo D, Bank A, Ward M, Malavasi F: Retrovirus-mediated transfer of the multidrug resistance gene into human hematopoietic progenitor cells. Br J Hematol 1994;88:318–324.
43   Cole SP, Bhardwaj G, Gerlach JH, Mackie JE, Grant CE, Almquist KC, Stewart AJ, Kurz EU, Duncan AM, Deeley RG: Overexpression of a transporter gene in a multidrug-resistant human lung cancer cell line [see comments]. Science 1992;258:1650–1654.
44   Grant CE, Valdimarsson G, Hipfner DR, Almquist KC, Cole SP, Deeley RG: Overexpression of multidrug resistance-associated protein (MRP) increases resistance to natural product drugs. Cancer Res 1994;54:357–361.
45   D'Hondt V, Caruso M, Bank A: Retroviral-mediated gene transfer of the multidrug resistance-associated protein cDNA protects cells from chemotherapeutic agents. Hum Gene Ther 1997;8: 1745–1751.
46   Machiels JP, D'Hondt V, Govaerts AS, Feyens AM, Bayat B, Bank A: Transfer and expression of multidrug resistance-associated protein cDNA into murine hematopoietic cells (abstract). Blood 1997;90(suppl):415b.
47   Hesdorffer C, Antman K, Bank A, Fetell M, Mears G: Clinical protocol: Human MDR gene transfer in patients with advanced cancer. Hum Gene Ther 1994;5:1151–1160.
48   Dunbar CE, Cottler-Fox M, O'Shaughnessy JA, Doren S, Carter C, Berenson R, Brown S, Moen RC, Greenblatt J, Stewart FM, Leitman SF, Wilson WH, Cowan K, Young NS, Nienhuis AW: Retrovirally marked CD34-enriched peripheral blood and bone marrow cells contribute to long-term engraftment after autologous transplantation. Blood 1995;85:3048–3057.
49   Hanania EG, Giles RE, Kavanagh J, Fu SQ, Ellerson D, Zu Z, Wang T, Su Y, Kudelka A, Rahman Z, Holmes F, Hortobagyi G, Claxton D, Bachier C, Thall P, Cheng S, Hester J, Ostrove JM, Bird RE, Chang A, Korbling M, Seong D, Cote R, Holzmayer T, Deisseroth AB, et al: Results of MDR-1 vector modification trial indicate that granulocyte/macrophage colony-forming unit cells do not contribute to posttransplant hematopoietic recovery following intensive systemic therapy. Proc Natl Acad Sci USA 1996;93:15346–15351.

50 Peters SO, Kittler EL, Ramshaw HS, Quesenberry PJ: Murine marrow cells expanded in culture with IL-3, IL-6, IL-11, and SCF acquire an engraftment defect in normal hosts. Exp Hematol 1995;23:461–469.
51 Peters SO, Kittler EL, Ramshaw HS, Quesenberry PJ: Ex vivo expansion of murine marrow cells with interleukin-3 (IL-3), IL-6, IL-11, and stem cell factor leads to impaired engraftment in irradiated hosts. Blood 1996;87:30–37.
52 Naldini L, Blomer U, Gallay P, Ory D, Mulligan R, Gage F, Verma I, Trono D: In vivo gene delivery and stable transduction of nondividing cells by a lentiviral vector. Science 1996;272:263–267.

Arthur Bank, MD, Columbia University, HHSC 1604,
701 West 168th Street, New York, NY 10032 (USA)
Tel. 212 305 4186, Fax 212 923 2090, E-Mail bank@cuccfa.ccc.columbia.edu

O^6-Benzylguanine-Resistant Mutant MGMT Genes Improve Hematopoietic Cell Tolerance to Alkylating Agents

Brian M. Davis, Omer N. Koç, Jane S. Reese, Stanton L. Gerson

Division of Hematology/Oncology, Case Western Reserve University, Cleveland, Ohio, USA

Mechanism of MGMT-Mediated Drug Resistance

The human methylguanine-DNA-methyltransferase (MGMT) gene encodes the 207-amino-acid DNA repair protein, O^6-alkylguanine DNA alkyltransferase (AGT), which removes alkyl lesions from the O^6 position of guanine [1]. The primary toxicity of methylating and chloroethylating agents classified as nitrosoureas, tetrazines and triazines occurs via alkylation at the O^6 position of guanine [2]. Although there are several known mechanisms of resistance to these agents, including glutathione-S-transferase [3], defects in mismatch repair [4] and polyamines [5], direct DNA repair of the alkyl adduct by AGT is the predominant repair mechanism associated with drug resistance [6–9].

The mechanism of action of alkyltransferase is unique among DNA repair enzymes. Repair occurs by covalent transfer of the O^6 adduct to a cysteine residue in the active site of the protein, an irreversible 'suicide' process which inactivates the transfer activity of the protein (fig. 1) [10]. Thus, one AGT protein stoichiometrically repairs one O^6-alkylguanine lesion, and a new AGT protein must be translated to repair additional lesions [11]. Although somewhat inefficient, the mechanism of action of AGT is not counterproductive for DNA repair; steady-state levels of AGT approximate 10,000 proteins per hematopoietic cell (determined from AGT activity/μg DNA) [6], which is more than sufficient to repair endogenous alkylation damage at O^6 of guanine, but not larger amounts of damage that are formed after treatment with alkylating agents. Expression of MGMT provides resistance to monofunctional methylating agents such as dacarbazine, temozolomide, streptozotocin and

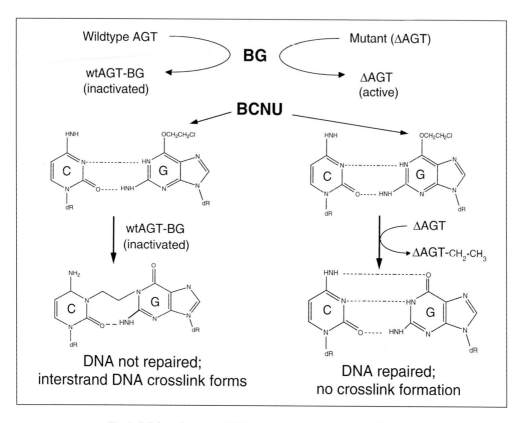

Fig. 1. BG inactivates wtAGT activity, preventing repair of the BCNU-derived chloroethyl lesion on the O^6 position of guanine. The lesion rearranges to form an N^{1}O^6-ethanoguanine intermediate, then forms a cytotoxic, covalent crosslink between the guanine and the antiparallel strand cytosine. Mutant AGT molecules retain their activity in the presence of BG, permitting AGT-mediated repair of the pre-crosslink lesion.

procarbazine [12, 13] as well as chloroethylating agents including 1,3-bis(2-chloroethyl)-1-nitrosourea (BCNU), 1-(4-amino-2-methyl-5-pyrimidinyl)methyl-3-(2-chloro)-3 nitrosourea (ACNU) and 3-cyclohexyl-1-chloroethyl-nitrosourea (CCNU) [6, 8].

O^6-chloroethylguanine lesions undergo rapid intramolecular rearrangement to the more stable N^{1}O^6-ethanoguanine [14]. This adduct is repairable by AGT, forming a covalent protein-DNA crosslink at the cysteine residue in the active site [2, 15]. If unrepaired, N^{1}O^6-ethanoguanine will form a highly toxic interstrand DNA crosslink with the antiparallel strand cytosine [14]. Cytotoxicity by methylating agents is due to mismatch repair complex recognition of the O^6-methylguanine:thymine repair base mispair, which is formed after one round of DNA replication, and induction of an aberrant repair process termed abortive mismatch repair, leading to multiple DNA strand breaks [16]. Interestingly, tumors deficient in mismatch repair activity have greater resistance to methylating agents [17–19], a paradox which appeared counterintuitive until the role of mismatch repair in methylating agent cytotoxicity was described.

Mammalian tissues display considerable variation in AGT expression. In humans, liver contains the highest AGT activity [20], while hematopoietic CD34+ cells possess the lowest activity [21, 22]. The low level of AGT activity in the bone marrow provides little protection from alkylating agents, which often results in severe myelosuppression after nitrosourea chemotherapy [22]. Thus, MGMT gene transfer has been proposed to increase bone marrow tolerance to alkylating agents, allowing dose escalation and more effective chemotherapeutic treatment of tumors.

The Limitations of Wild-Type MGMT Gene Transfer

We and others have shown the retroviral gene transfer of wild-type (wt) MGMT cDNA successfully confers nitrosourea resistance in vitro to primary murine hematopoietic progenitors [23–25] as well as human committed myeloid progenitors [26] and the more immature progenitor, the long-term culture-initiating cell (LTC-IC) [27], both of which are derived from CD34+ cells. Transplantation of wtMGMT-transduced murine hematopoietic progenitors into lethally irradiated mice resulted in increased resistance to BCNU in vivo and a 10- to 40-fold increase in AGT expression [23]. Maze et al. [25] reported that 92% MGMT transplanted mice compared to 53% of mock-transplanted mice survived weekly doses of 40 mg/kg over 5 weeks, a relatively high dose of drug considering the potential multiorgan cumulative toxicity observed with BCNU treatment. BCNU resistance was conferred to both myeloid and lymphoid [28] lineages in wtMGMT-transplanted mice. In contrast to mock-infected controls, mice infused with wtMGMT+ cells had increased leukocyte counts, platelet counts, and hematocrit, as well as a normal distribution of T cell subsets. The controls remained chronically pancytopenic [25], and had evidence of profound delayed myeloid and lymphoid suppression, consistent with the supposition that BCNU acts as a stem cell toxin. More than 6 months after treatment, the wtMGMT+ mice appeared to have maintained hematopoiesis with only a mild decrease in cellularity. Thus, transfer of wtMGMT generates enhanced BCNU resistance and a reduction in myelosuppression. Because AGT overexpression increases bone marrow tolerance to nitrosoureas, MGMT-transduced cells can be selected in vivo using these agents. Allay et al. [29] demonstrated that three cycles of in vivo BCNU administration at doses ranging from 15 to 40 mg/kg increased the proportion of bone-marrow-derived colony-forming units (CFU) containing the wtMGMT provirus in mice transplanted with wtMGMT-transduced cells from 37 to 90%. The BCNU IC_{50} in marrow-derived CFU obtained from mice transplanted with wtMGMT and given BCNU prior to sacrifice was 2-fold higher than that observed in untreated wtMGMT mice (19 vs. 10 μM) and 4-fold higher than mice transplanted with lacZ-transduced bone marrow progenitors (5 μM).

Although animal models of wtMGMT gene transfer demonstrated increased tolerance to nitrosourca chemotherapy, at the cellular level there was less than a 2-fold increase in BCNU resistance, which can be easily matched by drug-resistant tumors [30]. In fact, the expected result after transfer of wtMGMT would simply be to render the bone marrow as drug resistant as the tumor cells. Thus, the prognosis for the successful translation of wtMGMT gene transfer to a setting of cancer therapy is poor.

O^6-Benzylguanine-Resistant MGMT Mutants for Gene Transfer

The identification of mutant forms of AGT that are resistant to the potent AGT inhibitor, O^6-benzylguanine (BG), rekindled interest in MGMT gene transfer to improve hematopoietic cell chemotherapy tolerance. BG is a base analog that reacts with the active site of AGT, forming a covalent S-benzylcysteine moiety at the acceptor site cysteine [31], which results in the irreversible inactivation of AGT (fig. 2). BG has been shown to potentiate the cytotoxicity of alkylating agents to tumor cells in culture [32, 33] and in murine xenograft tumor experiments [34, 35], thereby improving the antitumor effect of these drugs. Phase 1 clinical trials using BG in combination with BCNU are currently under way at our institution [36, 37] and at the University of Chicago [38]. BG inactivation of tumor AGT has been found at doses as low as 10 mg/m^2, but complete inactivation requires much higher doses, 120 mg/m^2 [37]. In the course of the preclinical toxicity profile analysis of BG, increased toxicity to human hematopoietic progenitors was noted in vitro with BG and BCNU [22], and myelosuppression was observed with the combination of BG and BCNU in both mice [39] and dogs [40]. Thus, it was proposed that gene transfer of a BG-resistant mutant MGMT gene into hematopoietic progenitors, followed by treatment with alkylating agents in combination with BG, might simultaneously sensitize tumors, even those that are drug resistant due to high AGT expression, and protect the bone marrow from drug-induced myelosuppression.

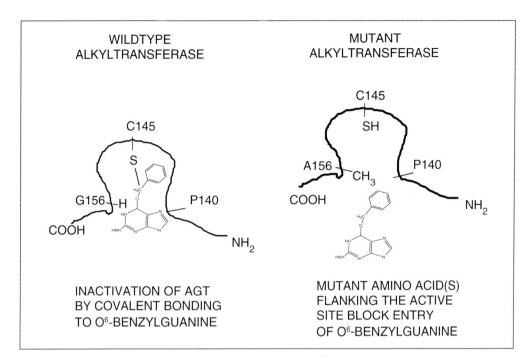

Fig. 2. The cysteine active site of wtAGT covalently binds to BG, permanently inactivating the protein. Mutant AGT proteins with amino acid substitutions in regions flanking the active site block the entry of BG, either by steric hindrance or by disrupting the natural electrostatic interactions after the endogenous residues were replaced with charged amino acids.

The bacterial homolog of the mammalian AGT gene, ada, is resistant to BG inactivation because of differences in the amino acid sequence flanking the active site [41]; as expected, retroviral gene transfer of ada resulted in improved resistance of murine hematopoietic progenitors to the BG and BCNU combination both in vitro and in vivo [42]. Harris et al. [42] noticed a narrow therapeutic window in mice transplanted with ada-transduced hematopoietic progenitors. While little toxicity was observed between mock and ada mice when treated with 30 mg/kg BG plus 10 mg/kg BCNU, both ada- and mock-transplanted mice were equally sensitive to 30 mg/kg BG and 15 mg/kg BCNU. Significant differential survival between ada- and mock-transplanted mice was observed only when treated with two doses of 30 mg/kg BG and 12.5 mg/kg BCNU. The effectiveness of ada gene transfer was limited by the relatively low DNA repair activity by bacterial Ada in mammalian cells, possibly due to the low proportion of the protein retained in the nucleus [43]. Additionally, expression of the bacterial protein in vivo might be immunogenic, further complicating ada gene transfer.

Since the amino acid sequence near the active site cysteine at residue 145 is conserved between the bacterial Ada and mammalian AGTs (IPCHRV for both), it was hypothesized that sequence differences in regions flanking the active site were responsible for the differential sensitivity to BG inactivation. This hypothesis was supported when the crystal structure of the Ada protein was resolved in 1994, and it was observed that the active site was buried within the protein [44]. Thus, it is believed that BG resistance is due to structural differences in Ada compared to the mammalian protein which sterically or electrostatically restricts the active site from accepting the bulky hydrophobic benzyl group of BG. Pegg and colleagues [45–47] designed human AGT proteins with increased BG resistance by replacing amino acids that flank the active site with residues similar to those found in Ada. Some mutants contained basic amino acid substitutions within the active site pocket, further restricting access of the benzyl group. Additional BG-resistant mutants were discovered by Loeb and colleagues, [48, 49], in experiments in which the natural DNA sequence between codons 150 and 172 of human AGT was replaced with random oligomers, which were then transfected into ada$^-$/ogt$^-$ GWR111 E. coli and selected for resistance to BG and MNNG.

In total, there are several mutant human AGT proteins that have been characterized as resistant to BG. These mutants include (but are not limited to): G156A [46], P140A [45], G160R [47] and the double mutant P140A/G156A [47] with ED_{50} values of inactivation by BG of 60, 5, 4 and >300 μM, respectively, compared to an ED_{50} of 0.25 μM for wtAGT. All retain the ability to remove methyl and chloroethyl groups from the O^6 position of guanine DNA and thus retain drug resistance activity in vitro. Because the maximum plasma concentration of BG and its active metabolite O^6-benzyl-8-oxo-guanine that can be achieved in humans is 10–15 μM at 120 mg/m^2 [37], the degree of BG resistance conferred by the most resistant mutants, most notably those that contain G156A, is believed to be sufficient for in vivo applications. BG-resistant mutant AGT molecules have multiple advantages over the bacterial Ada protein in mammalian cells, including sequences which encode an efficient nuclear localization signal and only minor amino acid substitutions relative to the foreign Ada protein which reduce the risk of immunogenicity, although this issue has yet to be addressed experimentally. However, there is mounting evidence that some of these mutant proteins may be less stable than the wt form. Newly described mutants including P140K, P138K/V139L/P140K (KLK) and P138M/V139L/P140K (MLK) appear to possess increased BG resistance [50] and improved stability over the other mutants and are currently under investigation by a number of groups.

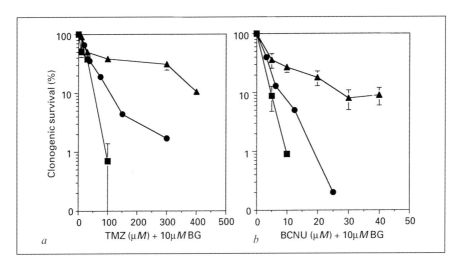

Fig. 3. SW480 tumor cells survive doses of BG and TMZ (*a*) or BG and BCNU (*b*) that are toxic to human CD34+ hematopoietic cells. However, transduction of CD34+ cells with ΔMGMT increases hematopoietic cell survival 5- to 8-fold untransduced CD34+ cells, and 3- to 4-fold above SW480 cells. ● = SW480 cells; ■ = untransduced CD34+ cells; ▲ = ΔMGMT-transduced CD34+ cells.

Experimental Models Using G156A MGMT (ΔMGMT)

Our laboratory has been studying gene transfer of the G156A mutant of MGMT, which we have designated ΔMGMT to distinguish this mutant from wtMGMT. We have subcloned the ΔMGMT cDNA into the MFG retroviral vector, which drives long-term, high-level expression in hematopoietic cells. Reese et al. [51] demonstrated that ΔMGMT transduction of human hematopoietic CD34+ cells provided 5.6-fold increased resistance to BCNU treatment in combination with 10 μM BG (with an IC$_{90}$ of 28 μM) relative to untransduced CD34+ cells (IC$_{90}$ of 5 μM), a much greater difference compared to the less than 2-fold increase observed after wtMGMT transduction after exposure to BCNU. Additionally, CFU derived from ΔMGMT-transduced human CD34+ cells demonstrated an 8-fold increase in resistance to the methylating agent temozolomide (TMZ) in combination with 10 μM BG (IC$_{90}$ of 370 μM vs. 52 μM) [52]. Whereas SW480 cells were more resistant to BG and BCNU or BG and TMZ than were untransduced CD34+ cells, ΔMGMT transduced CD34+ cells survived significantly higher doses of these drugs than did the tumor cells (which had a BCNU IC$_{90}$ of 9 μM and a TMZ IC$_{90}$ of 112 μM (fig. 3). We have also demonstrated efficient transduction and protection of earlier progenitors. ΔMGMT-transduced human LTC-IC were

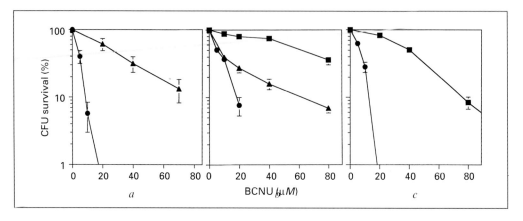

Fig. 4. BG and BCNU resistance in bone-marrow-derived CFU after transduction and before transplantation (*a*), from transplanted mice sacrificed 13 weeks after transplantation (*b*), and from transplanted mice sacrificed 23 weeks after transplantation (*c*). ● = lacZ-transduced CFU; ▲ = unselected ΔMGMT-transduced CFU; ■ = selected ΔMGMT-transduced CFU. Reprinted with permission from Davis et al. [54].

significantly more resistant to BG and BCNU compared with their untransduced counterparts [53]. However, both untransduced and ΔMGMT-transduced LTC-IC were similarly resistant to BG and TMZ [52]. Since TMZ toxicity is dependent upon DNA replication, the relatively quiescent LTC-IC may have sufficient time to repair the O^6-methyl lesions before entry into S-phase. Prolonged depletion of AGT by continual exposure to BG may be necessary to sensitize these cells to TMZ.

Murine committed myeloid hematopoietic progenitors acquire a similar degree of BG and BCNU and BG and TMZ resistance after transfer of ΔMGMT as do human CD34+ cells. The BG and BCNU IC_{50} in ΔMGMT-transduced CFU was 5- and 4-fold higher than in lacZ- and wtMGMT-transduced CFU, respectively [54]. Furthermore, 95% of mice transplanted with ΔMGMT-transduced primary hematopoietic progenitors survived repetitive treatment with 30 mg/kg BG and 10–20 mg/kg BCNU compared to only 23% of mock-transduced controls. This represents a greater differential survival using much lower doses of BCNU than required by the Williams group [25] in survival experiments using wtMGMT and BCNU alone, and ΔMGMT-transplanted mice could tolerate doses of BCNU that were lethal to mice transplanted with ada-transduced progenitors. Bone-marrow-derived CFU obtained from these mice were significantly protected from in vitro drug treatment with BG and BCNU 13 weeks and 23 weeks after transplantation, demonstrating efficient transduction of long-term repopulating progenitors (fig. 4). The BG and BCNU IC_{50} in CFU from BG- and BCNU-treated

ΔMGMT animals was 68 µM compared to 6.5 µM in untreated ΔMGMT and 6.2 µM in lacZ animals, representing a greater than 10-fold increase in drug resistance [54].

Preclinical Xenograft Studies

These experiments led to xenograft tumor models in which nude mice were infused with isogeneic ΔMGMT-transduced bone marrow without prior myeloablation and innoculated with BCNU-resistant human colon cancer SW480 cells, which overexpress AGT at levels significantly higher than hematopoietic cells (fig. 3). The xenograft studies mimic a potential clinical setting in which patients with solid tumors would be reinfused with autologous ΔMGMT-transduced hematopoietic progenitor cells after standard, nonmyeloablative doses of BG and BCNU. At 6-week intervals, patients would then receive cycles of BG and BCNU at standard chemotherapeutic doses. In the preclinical studies, tumor-bearing nude mice infused wth ΔMGMT transduced marrow survived 3–5 cycles of BG in combination with escalating doses of BCNU ranging from 10 to 25 mg/kg and displayed a significant delay in tumor growth compared to the tumor growth rate in untreated mice ($T_C - T_E = 60$ days) [55]. Mice receiving unmodified cells experienced a much greater degree of myelosuppression relative to their ΔMGMT-transduced counterparts. Therefore, there was an improved therapeutic index of multiple BG and BCNU treatments after ΔMGMT gene transfer, reducing myelotoxicity while maintaining significant tumor growth delay.

BG-Resistant MGMT Mutants Can Be Used for in vivo Selection

The dramatic increase in drug-resistant CFU after BG and BCNU administration implies that drug treatment enriches for transduced progenitors. BG- and BCNU-mediated selection for ΔMGMT-transduced progenitors is stronger than we observed for wtMGMT selection after BCNU alone. Davis et al. [54] demonstrated that at 13 weeks after transplantation, proviral ΔMGMT sequence was detected in 67% of CFU from untreated mice and increased to 100% of CFU from mice selected with one cycle of BG and BCNU [54]. Furthermore, expression of ΔAGT in the bone marrow and peripheral blood was tested by flow cytometry on permeabilized cells using a novel, well-defined method developed in our laboratory [56]. Using this technique, 30% of bone marrow cells from untreated ΔMGMT mice expressed ΔAGT above background, and one cycle of BG and BCNU increased the

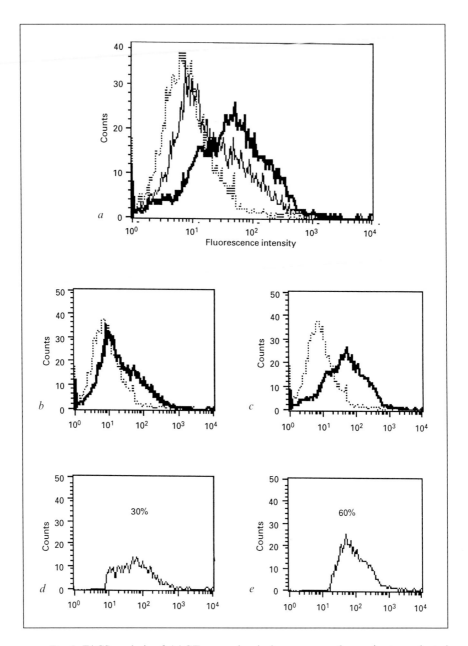

Fig. 5. FACS analysis of ΔAGT expression in bone marrow from mice transplanted with ΔMGMT- or lacZ-transduced progenitors 13 weeks after transplantation. *a* Comparison of expression in lacZ (dotted line), unselected ΔMGMT (solid line) and BG- and BCNU-selected ΔMGMT (bold line) bone marrow cells obtained from transplanted mice. *b* Overlay of lacZ and unselected ΔMGMT cells. *c* Resulting histogram after subtraction analysis

proportion of cells expressing ΔAGT to 60% (fig. 5) [54]. Fourteen percent of peripheral blood cells from untreated ΔMGMT animals expressed ΔAGT compared to 30% in peripheral blood cells from BG- and BCNU-treated ΔMGMT mice, an increase identical to the 2-fold enrichment seen in bone marrow.

Current protocols for retroviral-mediated gene transfer result in low level transduction into early hematopoietic progenitors and inefficient transgene expression in vivo. Therefore, in vivo selection for transduced cells may be necessary to achieve a level of genetically modified cells sufficient to generate a phenotypic change in an organism. Because the role of drug resistance genes in general is to protect cells from cytotoxic drug treatment, these genes intrinsically have the ability to act as dominant selectable markers during drug selection. Optimal enrichment would be achieved by selection for transduced stem cells, as these cells would self-renew, proliferate, and repopulate the bone marrow of the organism with genetically modified cells. However, certain drug resistance genes such as MDR1 and aldehyde dehydrogenase-1 (ALDH1) are endogenously expressed at high levels in early progenitors [57, 58], which may limit the ability to select for transduced cells carrying either of these genes. In contrast, with MGMT gene transfer, the low activity of AGT in early human hematopoietic progenitors make these cells naturally sensitive to BCNU, allowing for greater selection against untransduced cells and therefore should generate better enrichment for transduced progenitors. However, with a less than 2-fold relative resistance of wtMGMT-transduced vs. untransduced progenitors, the selection pressure for transduced early progenitors is modest and drug selection would take many cycles of treatment. On the other hand, BG depletion of endogenous AGT activity enhances the cytotoxic effect of BCNU such that relatively low levels of BCNU are required to mediate the cytotoxicity of untransduced hematopoietic and tumor cells. These doses would not be toxic to hematopoietic cells transduced with ΔMGMT, which have an approximate 20- to 100-fold survival advantage (based on studies mentioned below). Thus, the simultaneous sensitization of unmodified cells to BCNU by BG and the protection of transduced cells by ΔMGMT dramatically increases the overall selective advantage of transduced cells in response to these drugs. Similar results are likely to be achieved using recently described mutant forms of dihydrofolate reductase (DHFR) in combination with trimetrexate and thymidine transport inhibitors, which may be able to achieve a

demonstrating expression in 30% of the unselected ΔMGMT cell population. *d* Overlay of lacZ and BG- and BCNU-selected ΔMGMT cells. *e* Resulting histogram after subtraction analysis demonstrating expression in 60% of the selected ΔMGMT cell population. Reprinted with permission from Davis et al. [54].

similar sensitization of unmodified cells [59], although concern remains that unmodified early stem cells are not sensitized by this approach.

In order to achieve long-term genetic correction of hematopoietic cells, selection for transduced cells must occur at the stem cell level. Taxol (a selective agent used after MDR1 gene transfer) and methotrexate (used with DHFR gene transfer) are cytotoxic to cycling cells but not to quiescent stem cells [60–62]. Therefore, only short-term repopulation by transduced cells would be expected and there would be no enrichment at the stem cell level. This strategy would be appropriate if short-term expression of the transgene could elicit the therapeutic effect, e.g. cancer therapies, however these genes could not generate the long-term enrichment necessary for many gene replacement therapies. In contrast, BCNU-mediated DNA cross-linking is permanent if the pre-crosslink lesion is not repaired within 8 h [63], and the cytotoxicity will be manifested when quiescent hematopoietic early progenitor and stem cells enter the cell cycle. Thus, selection for ΔMGMT-transduced stem cells should occur under circumstances in which these cells can be genetically altered. Transduction of quiescent stem cells may be limited by the inability to efficiently target stem cells using retroviral vectors, which require progresssion through the cell cycle for stable integration of the provirus. Cytokine stimulation to force cells into cell cycle may result in progressive differentiation with each round of cell division, and may cause loss of the stem cell phenotype [64, 65]. Newly described lentiviral vectors can stably integrate into the genome of quiescent cells [66] and may overcome this limitation, allowing ΔMGMT gene transfer into stem cells and subsequent BG- and BCNU-mediated selection for the transduced stem cells both in vitro and in vivo. The small size of the ΔMGMT cDNA (621 bp) facilitates its inclusion in bicistronic vectors containing a therapeutic gene linked by an internal ribosome entry site, which allows for translation of two genes from one transcript. Because both genes are expressed from the same proviral promoter, selection for ΔMGMT expression would simultaneously select for high level expression of the therapeutic gene, theoretically resulting in permanent genetic correction of a deficient phenotype at the stem cell level.

Selection for ΔMGMT-Transduced Cells in Nonmyeloablated Mice

Infused ΔMGMT-transduced hematopoietic progenitors must also compete with endogenous stem cells. In the nonmyeloablated mouse, transduced progenitors do not contribute to long-term hematopoiesis unless large numbers of cells are transplanted [67, 68]. This suggests that transduced cells may be at a competitive disdvantage relative to endogenous stem cells. The mechanism of this disadvantage, if it exists, may be multifunctional including loss of the stem cell phenotype after cytokine stimulation and the low proportion of transduced stem cells present in the reinfused cell population. This suggests that in vivo selection for early hematopoietic progenitor and stem cells may be necessary to achieve long-term reconstitution with genetically modified cells. Therefore, we have studied the ability to select for ΔMGMT-transduced long-term repopulating progenitors infused into nonmyeloablated mice using BG and BCNU. Mice infused with a small number of hematopoietic progenitors, ranging from 1×10^6 to as low as 5×10^4 cells, and either given repetitive doses of BG and BCNU or left untreated, were analyzed for the presence of proviral ΔMGMT 24 to 30 weeks after infusion, at a time when all of the infused short-term repopulating cells should have been exhausted. In bone-marrow-derived CFU from mice left untreated, there was no evidence of the ΔMGMT proviral sequence. However, 15–97% of CFU from mice that were treated with BG and BCNU contained the ΔMGMT gene, in proportion to the number of cells that were initially infused and cycles of drug received [69]. These data, in combination with the lethally irradiated mouse transplant data reviewed above, suggest that there is an approximate 20- to 100-fold survival advantage to ΔMGMT-transduced hematopoietic progenitors compared to the less than 2-fold advantage to wtMGMT-transduced cells. After 3 cycles of treatment, this predicts hematopoietic progenitor cell enrichment of over 1,000-fold for ΔMGMT compared with at most 8-fold enrichment using wtMGMT. Thus, ΔMGMT appears to be a strong dominant selectable marker for long-term repopulating hematopoietic cells, clearly superior to wtMGMT and potentially better than other drug resistance genes.

Conclusion

In conclusion, ΔMGMT or other BG-resistant mutant MGMT genes, appear to offer a clear advantage for increasing the drug resistance of hematopoietic cells while sensitizing tumors during alkylating agent chemotherapy when coadministered with BG, compared to wtMGMT. Rather than simply making the marrow cells more resistant, the use of BG in combination with ΔMGMT gene transfer results in hematopoietic progenitors uniquely resistant to BCNU, thereby potentially increasing the therapeutic window of this drug combination. Mutant MGMT genes may also prove valuable or even essential for in vivo selection of transduced stem cells and appear to offer advantages over other drug resistance genes. ΔMGMT transduction of stem cells followed by treatment with BG and BCNU has the potential to generate selection at the stem cell level. The utility of this approach will be tested over the coming years, and if successful, may allow genetic modification at the stem cell level and life-long reconstitution of the bone marrow with genetically altered cells.

Acknowledgment

This work was supported in part by Public Health Service Grants RO1ES06288, PO1CA732062, RO1CA63193, P30CA43703, UO1CA75525 and MO1RR00080-35.

References

1 Pegg AE, Weist L, Foote RS, Mitra S, Perry W: Purification and properties of O^6-methylguanine-DNA transmethylase from rat liver. J Biol Chem 1983;258:2327–2333.
2 Gonzaga PE, Brent TP: Affinity purification and characterization of human O^6-alkylguanine-DNA alkyltransferase complexed with BCNU-treated, synthetic oligonucleotide. Nucleic Acid Res 1989; 17:6581–6590.
3 Hansson J, Edgren M, Ehrsson H, Ringborg U, Nilsson B: Effect of d,l-buthionine-S,R-sulfoximine on cytotoxicity and DNA cross-linking induced by bifunctional DNA-reactive cytostatic drugs in human melanoma cells. Cancer Res 1988;48:19–26.
4 Kawate H, Sakumi K, Tsuzuki T, Nakatsuru Y, Ishikawa T, Takahashi S, Takano H, Noda T, Sekiguchi M: Separation of killing and tumorigenic effects of an alkylating agent in mice defective in two of the DNA repair genes. Proc Natl Acad Sci USA 1998;95:5116–5120.
5 Seidenfeld J, Komar KA: Chemosensitization of cultured human carcinoma cells to 1,3-bis(2-chloroethyl)-1-nitrosourea by difluoromethylornithine-induced polyamine depletion. Cancer Res 1985;45:2132–2138.
6 Walker MC, Masters JRW, Margison GP: O^6alkyguanine-DNA alkyltransferase activity and nitrosourea sensitivity in human cancer cell lines. Br J Cancer 1992;66:840–843.
7 Gerson SL, Berger NA, Acre C, Petzold SJ, Willson JK: Modulation of nitrosourea resistance in human colon cancer by O^6-methylguanine. Biochem Pharmacol 1992;43:1101–1107.
8 Fujio C, Chang HR, Tsujimura T, Ishizaki K, Kitamura H, Ikenaga M: Hypersensitivity of human tumor xenografts lacking O^6-alkylguanine-DNA alkyltransferase to the anti-tumor agent 1-(4-amino-2-methyl-5-pyrimidinyl)methyl-3-(2-chloroethyl)-3-nitrosourea. Carcinogenesis 1989;10:351–356.
9 Day III RS, Ziolkowski CHJ, Scudiero DA, Meyer SA, Mattern MR: Human tumor cell strains defective in the repair of alkylation damage. Carcinogenesis 1980;1:21–32.
10 Pegg AE: Mammalian O^6-alkylguanine-DNA alkyltransferase: Regulation and importance in response to alkylating carcinogenic and therapeutic agents. Cancer Res 1990;50:6119–6129.
11 Bogden JM, Eastman A, Bresnick E: A system in mouse liver for the repair of O^6-methylguanine lesions in methylated DNA. Nucleic Acid Res 1981;9:3089–3103.
12 Foster BJ, Newell DR, Carmichael J: Preclinical phase I and pharmacokinetic studies with the dimethyl phenyltriazene CB310-277. Br J Cancer 1993;67:362–368.
13 Stevens MFG, Newlands ES: From triazines and triazenes to temozolomide. Eur J Cancer 1993; 29A:1045–1047.
14 Tong WP, Kirk MC, Ludlum DB: Formation of the cross-link 1-[N^3-deoxycyctidyl],2-[N^1-deoxyguanosinyl]-ethane in DNA treated with N,N^1-bis(2-chloroethyl)-N-nitrosourea. Cancer Res 1982;42:3102–3105.
15 Brent TP, Lestrud SO, Smith DG: Formation of DNA interstrand cross-links by the novel chloroethylating agent 2-chloroethyl(methylsulfonyl)-methandsulfonate: Suppression of O^6-alkylguanine-DNA alkyltransferase purified from human leukemic lymphoblasts. Cancer Res 1987;47:3384–3387.

16 Karran P, Macpherson P, Ceccotti S, Dogliotti E, Griffin S, Bignami M: O^6-methylguanine residues elicit DNA repair synthesis by human cell extracts. J Biol Chem 1993;286:15878–15886.

17 Branch P, Aquilina G, Bignami M, Karran P: Defective mismatch binding and a mutator phenotype in cells tolerant to DNA damage. Nature (Lond) 1993;362:652–654.

18 Branch P, Hampson R, Karran P: DNA mismatch repair binding defects, DNA damage tolerance and mutator phenotypes in human colorectal carcinoma cell lines. Cancer Res 1995;55:2304–2309.

19 Liu L, Markowitz S, Gerson SL: Mismatch repair mutations override alkyltransferase in conferring resistance to temozolomide but not to 1,3-bis(2-chloroethyl)nitrosourea. Cancer Res 1996;56:5375–5379.

20 Gerson SL, Trey JE, Miller K, Berger NA: Comparison of O^6 alkylguanine-DNA alkyltransferase activity based on cellular DNA content in human, rat and mouse tissues. Carcinogenesis 1986;7:745–749.

21 Gerson SL, Miller K, Berger NA: Comparison of O^6-alkyltransferase activity in human myeloid cells. J Clin Invest 1985;76:2106–2114.

22 Gerson SL, Phillips WP, Kastan M, Dumenco LL, Donovan C; Human CD34 hematopoietic progenitors have low, cytokine-unresponsive O^6-alkylguanine-DNA alkyltransferase and are sensitive to O^6-benzylguanine plus BCNU. Blood 1996;88:1649–1655.

23 Allay JA, Dumenco LL, Koc ON, Liu L, Gerson SL: Retroviral transduction and expression of the human alkyltransferase cDNA provides nitrosourea resistance to hematopoietic cells. Blood 1995;85:3342–3351.

24 Moritz T, Mackay W, Glassner BJ, Williams DA, Samson L: Retrovirus-mediated expression of a DNA repair protein in bone marrow protects hematopoietic cells from nitrosourea-induced toxicity and in vitro and in vivo. Cancer Res 1995;55:2608–2614.

25 Maze R, Carney JP, Kelley MR, Glassner BJ, Williams DA, Samson L: Increasing DNA repair methyltransferase levels via bone marrow stem cell transduction rescues mice from the toxic effects of 1,3-bis(2-chloroethyl)-1-nitrosourea, a chemotherapeutic alkylating agent. Proc Natl Acad Sci USA 1996;93:206–210.

26 Allay JA, Koc ON, Davis BM, Gerson S: Retroviral-mediated gene transduction of the human alkyltransferase cDNA confers nitrosourea resistance to human hematopoietic progenitors. Clin Cancer Res 1996;2:1353–1359.

27 Jelinek J, Fairbairn LJ, Dexter TM, Rafferty JA, Stocking C, Ostertag W, Margison GP: Long-term protection of hematopoiesis against the cytotoxic effects of multiple doses of nitrosourea by retrovirus-mediated expression of human O^6-alkylguanine-DNA-alkyltransferase. Blood 1996;87:1957–1961.

28 Maze R, Kaput R, Kelley MR, Hansen WK, Oh SY, Williams DA: Reversal of 1,3-bis(2-chloroethyl)-1-nitrosourea-induced severe immunodeficiency by transduction of murine long-lived hematopoietic progenitor cells using O^6-methylguanine DNA methyltransferase complementary DNA. J Immunol 1997;158:1006–1013.

29 Allay JA, Davis BM, Gerson SL: Human alkyltransferase-transduced murine myeloid progenitors are enriched in vivo by BCNU treatment of transplanted mice. Exp Hematol 1997;25:1069–1076.

30 Phillips WP Jr, Willson JKV, Markowitz SD, Zborowska E, Zaidi NH, Liu L, Gordon NH, Gerson SL: MGMT transfectants of a 1,3-bis(2-chloroethyl)-1-nitrosourea (BCNU)-sensitive colon cancer cell line selectivity repopluate heterogeneous MGMT+/MGMT− xenografts after BCNU and O^6-benzylguanine plus BCNU. Cancer Res 1997;57:4817–4823.

31 Pegg AE, Boosalis M, Samson L: Mechanism of inactivation of human O^6-alkylguanine-DNA alkyltransferase by O^6-benzylguanine. Biochemistry 1993;32:11998–20006.

32 Dolan ME, Mitchell RB, Mummert C, Moschel RC, Pegg AE: Effect of O^6-benzylguanine analogues on sensitivity of human tumor cells to the cytotoxic effects of alkylating agents. Cancer Res 1991;51:3367–3372.

33 Dolan ME, Pegg AE, Moschel RC, Grindey GB: Effect of O^6-benzylguanine on the sensitivity of human colon tumor xenografts to 1,3-bis(2-chloroethyl)-1-nitrosourea (BCNU). Biochem Pharmacol 1993;46:285–290.

34 Friedman HS, Dolan ME, Moschel RC, Pegg AE, Felker GM, Rich J, Bigner DD, Schold SC: Enhancement of nitrosourea activity in medullablastoma and glioblastoma multiforme. J Natl Cancer Inst 1992;84:1926–1931.

35 Michell RB, Moschel RC, Dolan ME: Effect of O^6-benzylguanine on the sensitivity of human tumor xenografts to 1,3-bis(2-chloroethyl)-1 nitrosourea and on DNA interstrand crosslink formation. Cancer Res 1992;52:1171.
36 Spiro TP, Willson JKV, Haaga J, Hoppel CL, Liu L, Majka S, Gerson SL: O^6-benzylguanine and BCNU: Establishing the biochemical modulatory dose in tumor tissue for O^6-alkyguanine DNA alkyltransferase directed DNA repair. Proc ASCO 1996;15:177.
37 Spiro TP, Gerson SL, Hoppel CL, Liu L, Schupp JE, Majka S, Hagga J, Willson JKV: O^6-benzylguanine totally depletes alkylguanine DNA alkyltransferase in tumor tissue: A phase I pharmacokinetic/pharmacodynamic study. Proc ASCO 1998;17:212.
38 Dolan ME, Roy SK, Paras P, Fasanmade AA, Schilsky RL, Ratain MJ: O^6-benzylguanine in humans: Metabolic, pharmacokinetic and pharmacodynamic findings. J Clin Oncol 1998;16:1803–1810.
39 Chinnasamy N, Rafferty J, Hickson I, Ashby J, Tinwell H, Margison G, Dexter M, Fairbairn L: O^6-benzylguanine potentiates in in vivo toxicity and clastogenicity of temozolomide and BCNU in mouse bone marrow. Blood 1997;89:1566–1573.
40 Page JG, Giles HD, Phillips WP, Gerson SL, Smith AC, Tomaszweski JE: Preclinical toxicology study of O^6-benzylguanine (NSC-637037) and BCNU (Carmustine, NSC-409962) in male and female beagle dogs. Proc AACR 1994;35:328.
41 Elder R, Marginson G, Rafferty J: Differential inactivation of mammalian and *Escherichia coli* O^6-alkylguanine DNA alkyltransferases by O^6-benzylguanine. Biochem J 1994;298:231–235.
42 Harris LC, Marathi UK, Edwards CC, Houghton PJ, Srivastava DK, Vanin EF, Sorentino BJ, Brent TP: Retroviral transfer of a bacterial alkyltransferase gene into murine bone marrow protects against chloroethylnitrosourea cytotoxicity. Clin Cancer Res 1995;1:1359–1365.
43 Dumenco LL, Warman B, Hatzoglou MK, Lim IL, Abboud SL, Gerson SL: Increase in nitrosourea resistance in mammalian cells by retrovirally mediated gene transfer of bacterial O^6alkylguanine-DNA alkyltransferase. Cancer Res 1989;49:6044–6051.
44 Moore MH, Gulbis JM, Dodson EJ, Demple B, Moody PC: Crystal structure of a suicidal DNA repair protein: The Ada O^6-methylguanine-DNA methyltransferase from *E. coli*. EMBO J 1994;13:1495–1501.
45 Crone TM, Pegg AE: A single amino acid change in human O^6-alkylguanine-DNA alkyltransferase decreasing sensitivity to inactivation by O^6-benzylguanine. Cancer Res 1993;53:4750–4753.
46 Crone TM, Goodtzova K, Edara S, Pegg AE: Mutations in human O^6-alkylguanine-DNA alkyltransferase imparting resistance to O^6-benzylguanine. Cancer Res 1994;54:6221–6227.
47 Edara S, Kanugula S, Goodtzova K, Pegg AE: Resistance of the human O^6 alkylguanine DNA alkyltransferase containing arginine at codon 160 to inactivation by O^6 benzylguanine. Cancer Res 1996;56:5571–5575.
48 Christians FC, Dawson BJ, Coates MM, Loeb LA: Creation of human alkyltransferases resistant to O^6-benzylguanine. Cancer Res 1997;57:2007–2012.
49 Encell LP, Coates MM, Loeb LA: Engineering human DNA alkyltransferases for gene therapy using random sequence mutagenesis. Cancer Res 1998;58:1013–1020.
50 Xu-Welliver M, Pegg AE: Isolation of human O^6 alkylguanine DNA alkyltransferase (AGT) mutants highly resistant to O^6 benzylguanine (BG). Cancer Res 1998;58:1936–1945.
51 Reese JS, Koç ON, Lee KM, Liu L, Allay JA, Phillips WP, Gerson SL: Retroviral transduction of a mutant MGMT into human CD34 cells confers resistance to O^6-benzylguanine plus BCNU. Proc Natl Acad Sci USA 1996;93:14088–14093.
52 Reese JS, Davis BM, Liu L, Gerson SL: Simultaneous protection of G156A MGMT transduced hematopoietic progenitors and sensitization of tumor cells after O^6 benzylguanine and temozolomide. Clin Cancer Res, in press.
53 Koç ON, Reese JS, Szekely EM, Gerson SL: Human long term culture initiating cells are sensitive to BG and BCNU and protected after ΔMGMT gene transfer. Cancer Gene Ther, in press.
54 Davis BM, Reese JS, Koç ON, Lee K, Schupp JE, Gerson SL: Selection for G156A O^6-methylguanine DNA methyltransferase gene-transduced hematopoietic progenitors and protection from lethality in mice treated with O^6-benzylguanine and 1,3-bis(2-chloroethyl)-1-nitrosourea. Cancer Res 1997;57:5093–5099.

55 Koç ON, Davis BM, Reese JS, Liu L, Gerson SL: ΔMGMT transduced bone marrow infusion increases tolerance to O^6-benzylguanine and BCNU and allows intensive therapy of BCNU resistant xenografts in nude mice. Hum Gene Ther, in press.

56 Liu L, Lee K, Schupp JS, Koç ON, Gerson SL: Heterogeneity of O^6-alkylguanine DNA alkyltransferase measured by flow cytometric analysis in blood and bone marrow mononuclear cells. Clin Cancer Res 1998;4:475–481.

57 Drach D, Zhao S, Mahadevia R, Gattringer C, Huber H, Andreeff M: Subpopulations of normal peripheral blood and bone marrow cells express a functional multidrug resistant phenotype. Blood 1992;80:2729–2734.

58 Kastan MB, Schlaffer E, Russo JE, Colvin OM, Civin CI, Hilton J: Direct demonstration of elevated aldehyde dehydrogenase in human hematopoietic progenitor cells. Blood 1990;75:1947–1950.

59 Allay J, Spencer H, Wilkinson S, Belt J, Blakley R, Sorrentino B: Sensitization of hematopoietic stem and progenitor cells to trimetrexate using nucleoside transport inhibitors. Blood 1997;90:3546–3554.

60 Blau CA, Neff T, Papayannopoulou T: Cytokine prestimulation as a gene therapy strategy: Implications for using the MDR1 gene as a dominant selectable marker. Blood 1997;89:146–154.

61 Bertolini F, Battaglia M, Lanza A, Gibelli N, Palermo B, Pavesi L, Capriotti M, Robustelli della Cuna G: Multilineage long-term engraftment potential of drug resistant hematopoietic progenitors. Blood 1997;90:3027–3036.

62 Blau CA, Neff T, Papayannopoulou T: The hematologic effects of folate analogs: Implications for using the dihydrofolate reductase gene for in vivo selection. Hum Gene Ther 1996;7:2069–2078.

63 Gonzaga PE, Potter PM, Niu TQ, Yu D, Ludlum DB, Rafferty J, Margison GP, Brent TP: Identification of the cross-link between human O^6-methylguanine-DNA methyltransferase and chloroethylnitrosourea-treated DNA. Cancer Res 1992;52:6052–6058.

64 Peters SO, Kittler ELW, Ramshaw HS, Quesenberry PJ: Murine marrow cells expanded in culture with IL-3, IL-6, IL-11, and SCF acquire an engraftment defect in normal hosts. Exp Hematol 1995;23:461–469.

65 Kittler EL, Peters SO, Crittenden RB, Debatis ME, Ramshaw HS, Stewart FM, Quesenberry PJ: Cytokine-facilitated transduction leads to low-level engraftment in nonmyeloablated hosts. Blood 1997;90:865–872.

66 Naldini L, Blomer U, Gallay P, Ory D, Mulligan R, Gage FH, Verma IM, Trono D: In vivo gene delivery and stable transduction of non-dividing cells by a lentiviral vector. Science 1996;272:263–267.

67 Stewart FM, Crittenden RB, Lowry PA, Pearson-White S, Quesenberry PJ: Long-term engraftment of normal and post-5-fluorouracil murine marrow into normal non-myeloablated mice. Blood 1993;81:2566–2571.

68 Rao SS, Peters SO, Crittenden RB, Stewart FM, Ramshaw HS, Quesenberry PJ: Stem cell transplantation in the normal nonmyeloablated host: Relationship between cell dose, schedule, and engraftment. Exp Hematol 1997;25:114–121.

69 Davis BM, Koç ON, Gerson SL: Detection of long-term hematopoiesis by G156A MGMT transduced progenitors in non-myeloablated mice after enrichment with O^6-benzylguanine and BCNU. Blood 1997;90:554a.

Stanton L. Gerson, MD, Chair, Division of Hematology/Oncology,
Case Western Reserve University, 10900 Euclid Ave. BRB 3-West, Cleveland, OH 44106 (USA)
Tel. 216 368 1176, Fax 216 368 1166, E-Mail slg5@po.cwru.edu

Use of Variants of Dihydrofolate Reductase in Gene Transfer to Produce Resistance to Methotrexate and Trimetrexate[1]

J.R. Bertino, S.C. Zhao, S. Mineishi, E.A. Ercikan-Abali, D. Banerjee

Molecular Pharmacology and Therapeutics Program, Sloan Kettering Institute for Cancer Research, New York, N.Y., USA

Methotrexate (MTX), a potent inhibitor of dihydrofolate reductase (DHFR), that supplanted aminopterin, has enjoyed widespread use as an anticancer agent, an immunosuppressive agent, and to treat psoriasis and rheumatoid arthritis. To treat patients with malignancies, MTX is usually administered in doses that produce some toxicity to bone marrow or the gastrointestinal tract, in order to exert maximum antitumor effects. Leucovorin, an antidote for MTX toxicity, when used 24–48 h after MTX allows high doses of MTX to be administered, with an improvement in the therapeutic index in the treatment of patients with acute lymphocytic leukemia or lymphoma [1, 2]. In contrast, treatment of patients with non-neoplastic diseases, e.g. rheumatoid arthritis and psoriasis, only requires low-dose intermittent dosing (usually weekly) of this drug, at levels that do not ordinarily produce toxicity [3, 4]. However, in patients who are elderly or have poor diets, or with impaired renal function, severe and even fatal toxicity has been observed, even with these low doses.

The finding that cellular resistance to MTX may occasionally be due to mutations in the DHFR gene, whose product is the target for MTX, has prompted studies to utilize these mutant cDNAs and other mutations generated in the laboratory, i.e. by mutagenesis or site-directed mutagenesis, for use in gene therapy approaches with the objective to protect hematopoietic progenitor cells from MTX and TMTX toxicity.

[1] Supported by NIH grant CA 59350.

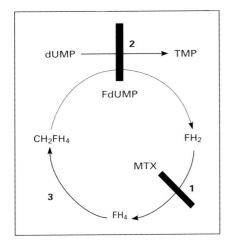

Fig. 1. The thymidylate cycle. dUMP = Deoxyuridylate; TMP = thymidylate; FdUMP = 5-fluorodeoxyuridylate; CH_2FH_4 = 5–10 methylene FH_4; FH_2 = dihydrofolate; 1 = dihydrofolate reductase; 2 = thymidylate synthase; 3 = serine hydroxymethylase.

The advantages of using DHFR mutant cDNAs for this purpose are the extensive use of MTX in the treatment of both neoplastic and non-neoplastic diseases, a detailed understanding of its mechanism of action and clinical pharmacology, the relatively small size of the cDNA for DHFR (less than 1.1 kb), thus allowing this cDNA along with other cDNAs to be easily accommodated into retroviral vectors, and the availability of mutant forms of DHFR that give rise to high levels of resistance to MTX and trimetrexate (TMTX), and yet are catalytically active.

In this manuscript, various mutant forms of DHFR, and retroviral constructs containing these cDNAs are described. Preclinical studies are reviewed, and clinical studies in the planning stage are outlined.

Mechanism of Action of Methotrexate and Resistance

DHFR catalyzes the reduction of dihydrofolate, a product in the synthesis of thymidylate to tetrahydrofolate (FH_4). Methylene FH_4, the cofactor for thymidylate synthase (fig. 1), is then generated via the enzyme serine hydroxymethylase, allowing thymidylate synthesis to continue. MTX and TMTX are powerful inhibitors of mammalian DHFRs, with K_i values in the picomolar range [4, 5]. Studies of acquired resistance to MTX in cell lines, and in murine tumors propagated in vivo, have shown that overexpression of this enzyme, due to amplification of the DHFR gene, and impaired uptake, due to decreased expression or mutations in the reduced folate carrier, are the major changes in the tumor cell phenotype that enable cells to survive the lethal effects of

Table 1. MTX-resistant cell lines with mutations in DHFR

Cell line	Mutation	Reference
3T6	L22R	11, 12
HCT8	F31S	13
L5178Y	F31N	14, 15
CHO	L22F	16, 17
L1210	G15W	18

3T6 = Mouse fibroblast cell line; HCT8 = human colon adenocarcinoma; L5178Y = mouse leukemia cell line; CHO = Chinese hamster ovary cell line; L1210 = mouse leukemia cell line.

this drug [1, 2, 7]. Less commonly, mutations in the DHFR that weaken the binding of MTX to this protein occur in resistant sublines, usually with overexpression of this DHFR protein, and lead to a high level of resistance to this drug. In the clinic, limited information is available concerning acquired resistance mechanisms to MTX. Recently leukemia blasts from patients with acute lymphocytic leukemia that were clinically resistant to MTX were shown to have impaired uptake of MTX, mainly due to decreased RFC expression, and less commonly, low-level DHFR gene amplification [7]. Thus far, alterations in DHFR leading to decreased MTX binding have not been found in tumor cells from patients resistant to MTX [7, 8].

Variant Forms of Dihydrofolate Reductase That Are Useful for Gene Transfer and Protection from Methotrexate and Trimetrexate Toxicity

Five different mutations have been described that are associated with cell lines selected for resistance to MTX (table 1). Of interest, all but the G15W mutant involve active site hydrophobic amino acids, L22 and F31. In a study using EMS mutagenesis and selection with trimetrexate, the only mutant form of DHFR generated was also at amino acid 31 (F31S) [9]. The knowledge of the crystal structure of human DHFR [10], and the amino acids involved in the binding of substrates and cofactors to this protein, as well as the information provided by the naturally occurring mutations in DHFR (table 1), have led several laboratories to generate new mutations by site-directed mutagenesis, with the goal of generating variants that provide a high level of resistance

Table 2. DHFR mutants useful for generating resistance to MTX: catalytic activity and inhibition by antifolates

Enzyme	$K_m(H_2$ folate) μM	K_{cat} S^{-1}	K_{cat}/K_m	K_i nM MTX	K_i nM TMTX	$K_i(K_{cat}/K_m)$ $S^{-1} \times 10^3$	Increase	Ref.
wt	0.08	12.7	159	0.0012	0.013	0.19	×1	19
L22F	0.16	7.4	46	0.11	0.083	4.9	×26	19
L22R	1.6	0.045	0.03	4.6	1,820	0.085	×0.5	19
L22Y	0.15	1.5	10	1.98	2.51	20	×100	19
F31S	0.44	7.0	16	0.24	0.24	3.8	×20	20
F31R	0.62	0.93	1.5	7.2	–	10.8	×56	21
F34S	8.77	1.43	0.16	210	–	33.6	×170	20
L22F, F31G	0.33	1.3	3.8	29	14	90.9	×478	22
L22F, F31S	0.44	1.6	3.6	26	19	96.5	×508	22
L22Y, F31G	0.35	0.5	1.4	150	910	160	×842	22
L22Y, F31S	0.40	1.3	3.3	42	400	216	×1140	22

[19–23]. Mutations at several hydrophobic active-site amino acids have been generated and explored in detail, in particular L22 and F31, with generation of variants that have desirable attributes for use in gene transfer, namely a high level of MTX resistance, good catalytic activity, and intracellular stability [5, 19–22]. Table 2 lists the variants for which data on catalytic activity and inhibition by MTX and TMTX have led to their use in gene transfer studies. A large number of other mutations have been generated, but limited information is available on some of these variants, either because of low catalytic activity or lack of protection against MTX inhibition [23–29]. The L22R mutant, the first mutant of DHFR described to give rise to MTX resistance [11, 12], provides a high level of resistance to MTX, and especially TMTX; however, the catalytic activity of this enzyme is very poor, and multiple copies are necessary to generate sufficient FH_4 to allow thymidylate synthesis. It is of interest that this mutant arose in a cell line that first amplified nonmutated DHFR, then with further selection the mutated and amplified L22R mutant was found [11]. This mutant has limited value for retroviral transduction of normal hematopoietic precursors, as only one or two copies are transferred via this method, and in vivo treatment with MTX or TMTX does not lead to amplification of normal cells [30, 31]. In our own experience, mice receiving marrow transduced with a retroviral vector containing the F31S variant and treated with MTX over a long period of time did not show an increase in the gene copy number of DHFR in hematopoietic progenitors (CFU-S). In addition, transfection of human CD34+ cells with a retrovirus containing this DHFR cDNA after long-term culture in the presence of MTX did not lead to amplification of the DHFR gene (either endogenous or transfected cDNA). These results are in accord with clinical observations of patients with psoriasis or rheumatoid arthritis treated with MTX for extended periods in that resistance to this drug is not observed. Both the L22Y and the F31 variants have been used for the transduction of mouse marrow progenitors, as these variants produce a reasonably high level of resistance and also have good catalytic properties (table 2). More recently, our laboratory has generated double mutants of DHFR, namely L22Y/F31S, L22Y/F31G, L22F/31G and L22F/31S (table 2). These double mutants have K_is for MTX and TMTX that are 5 logs greater than for wt DHFR [22]. Although there is some loss of catalytic activity, if one uses the combination of kinetic parameters as suggested by Blakley [5], $K_i \cdot (K_{cat}/K_m)$ is 4–5 times greater for the double mutants as compared to the best single mutants. When 3T3 cells were infected with a retroviral construct to compare the level of resistance generated by the single mutant F31S to the double mutant L22F/31S, the double mutant provided greater protection to high levels of MTX.

The Mouse as a Model for Transduction of Hematopoietic Precursors

Mice have been utilized primarily as a model system for purposes of proving principles and testing new vectors and concepts. Two major differences between the mouse model and humans have emerged that must be considered in relating these studies to patients. Mouse marrow progenitors have proven relatively easy to transfect, and long-term expression at relatively high levels are observed, as compared to studies in patients (using MDR cDNA). In addition, mice have high levels of folates and thymidine, and thus inhibitors of thymidine salvage may be necessary for ablation of early marrow progenitors, when using MTX or inhibitors of thymidylate synthase as selecting agents.

Despite these limitations, studies of transfection of mouse marrow progenitors in vitro, as well as in vivo studies after ex vivo transfections, have provided guidelines with regard to optimal vector use, optimal variants of DHFR, and dose schedules of antifolates, that are of value for planning human studies. Larger species, such as dogs and monkeys may be of more value in prediction of outcomes of gene therapy in patients, but the costs involved, the limited availability of these species, and limited availability of species-specific cytokines, has hampered the use of these species for preclinical studies. Calcium-phosphate-mediated gene transfer into mouse bone marrow using genomic DNA from the 3T6 line, where the original Arg22 mutation was characterized,

Table 3. Survival of recipients transplanted with bone marrow cells infected with DHFR^r or Neo^r virus and subsequently treated with MTX for 4 weeks

Group infected with	MTX	Survival
DHFR^r gene	–	12/12
DHFR^r gene	+	11/12
Neo^r gene (control)	+	0/12

Reproduced with permission from Williams et al. [35].

was the first report that such transfer of a mutant DHFR could result in resistance of the bone marrow to MTX [33]. This genetic transformation of murine bone marrow cells to MTX resistance was further improved and extended [34]. The next improvement in the field came with the retrovirus-mediated gene transfer of the Arg22 mutant DHFR cDNA into murine bone marrow cells [35]. The major problem with calcium-phosphate-mediated gene transfer was the relative inefficiency of the transformation process, while the retrovirus-mediated gene transfer proved to be far superior in this regard [35]. The first gene transfer of a mutant DHFR cDNA using a retroviral vector used the zipDHFR vector which utilizes the long-terminal repeat of the Moloney leukemia virus as promoter and does not carry any other marker. Transduction of murine bone marrow by coculture of bone marrow cells with virus-producing cells and subsequent transplantation of transduced marrow led to successful protection of the recipient marrow from MTX toxicity. The actual infection of an early stem cell was further demonstrated by serial transplantation of the marrow to secondary recipients who were also shown to be protected similarly [35, 36] (fig. 2).

In an effort to pursue the transfer of MTX resistance to bone marrow cells using better DHFR mutants, we initiated studies using the Ser31 human mutant DHFR cDNA as our initial studies indicated that this mutant was more efficient than the Arg22 mutant [37]. The Ser31 mutant human DHFR cDNA was cloned into the Moloney leukemia virus derived N2A vectors which carried the neo^r gene as another selectable marker. The vectors used were the modified double-copy vectors where the gene of interest is cloned into the 3'LTR of the vector, so that during its cycle the virus will produce a copy of the inserted gene in the 5'LTR as the virus uses the 3'LTR as a template to make the 5'LTR in the next cycle [38, 39]. Thus the vector ends up with two copies of the inserted gene, hence the name 'double-copy vector'. Further-

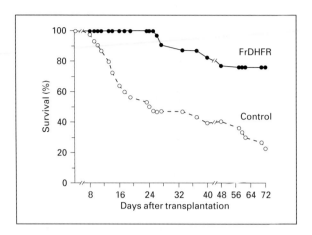

Fig. 2. Survival plot of animals after secondary transplantation of transduced bone marrow cells. Primary animals were harvested at 8 weeks posttransplantation, and 5×10^6 nucleated cells were infused into irradiated secondary recipients. MTX was given to secondary recipients as described in Materials and Methods. Results represent a total of 30 animals in each group. The difference in survival is significant ($p < 0.000003$) when analyzed by the log rank test [25, 26]. Reproduced with permission from Corey et al. [36].

more, the inserted mutant DHFR cDNA is placed under the transcriptional control of various internal promoters in reverse orientation to the LTR promoter. This was done with a view to circumvent the possibility of reduced transcription off an internal promoter in the same orientation as the viral LTR promoter. Of the five different promoters tested in various cell systems, it was observed that the SV40 and the human ADA promoters were the most efficient in bone marrow cells [39]. It was therefore reasonable to carry out comparative studies both in vitro and in vivo with the two promoters and the two DHFR mutants the Arg22 and the Ser31 DHFR cDNA. It was observed that the Ser31 DHFR cDNA was able to protect mice from methotrexate toxicity, and both secondary and tertiary recipients were also protected, indicating that this protection was due to transduction of an early progenitor or stem cell [40–42].

Direct evidence of gene transfer was obtained by polymerase chain reaction (PCR) amplification of the transfected neo^r sequences and hybridization of Southern blots with neo^r specific probes after restriction digestion of genomic DNA obtained from day 12 colony-forming-units-spleen (CFU-S) colonies. Long-term expression of the introduced gene was demonstrated by serial transplantation from primary to secondary to tertiary murine recipients. In vivo enrichment of the transduced mutant human DHFR was also shown by sequential increases in percentage of MTX resistant colony-forming-units-granulocyte macrophage (CFU-GM) colonies from primary to secondary murine recipients. Moreover, direct evidence for the presence of the mutant DHFR gene was obtained by sequencing the PCR-amplified mutant DHFR sequences using primers specific for the SV40 promoter and the transferred DHFR gene. Various dose schedules were also employed in order to establish a more clinically relevant situation. After transduction and bone marrow transplantation with the mutant DHFR, the recipients were treated with 5 mg of MTX per kilogram body weight for the first week and 10 mg of MTX per kilogram per week for the next 6 weeks. The control animals all died within the first 3 weeks while the experimental animals survived the MTX treatment [40].

In another experimental design, transduced animals were not treated with MTX for the first 4 weeks after bone marrow transplantation but were then treated with high-dose MTX (200 mg/kg for 2 weeks and then 400 mg/kg for the next 6 weeks). This delayed high-dose selection allowed the recipients to recover completely from the irradiation and bone marrow transplantation before being subjected to MTX treatment. The 4-week delay allowed selection with a much larger dose of MTX. The delayed high-dose selection with MTX resulted in the deaths of all control animals within 4 weeks while more than 80% of experimental animals survived the treatment.

Results of studies from Quesenberry's laboratory, as well as our own (unpubl. observ.), indicate that marrow ablation may not be necessary to obtain repopulation of marrow with cells transfected ex vivo, if larger numbers of hematopoietic progenitor cells are transduced [43]. These results raise the possibility that it may be possible to harvest autologous peripheral blood stem cells from patients after G-CSF treatment, and after infection with retrovirus, return the marrow to the unirradiated patient and utilize MTX or TMTX to allow selection and expression of progenitors expressing mutant DHFR (or any other drug-resistant gene). If successful, even patients with nonneoplastic disease undergoing long-term treatment with MTX may be candidates for this approach, allowing safe use of this drug, even with increased doses. It has been pointed out that gastrointestinal rather than hematologic toxicity may be dose limiting when patients receive MTX. However, in the posttransplantation period, this may not be the case, as evidenced by the poor hematologic tolerance of patients to this drug after receiving an allograft, when this drug is used to prevent graft-versus-host disease. Therefore, if higher doses of MTX are tolerated posttransplantation, after transfection of marrow cells with a mutant DHFR cDNA, this procedure may allow larger doses of MTX to be used safely in this setting (as an immunosuppressive agent), as well as in other situations as an antitumor agent, e.g. breast cancer, lymphoma. The safe use of this drug alone or in combination with another drug whose toxicity is also diminished by the use of a vector that contains a second cDNA that provides resistance, in particular in lymphoma or breast cancer patients at high risk of relapse, could provide additional tumor cell kill that may possibly increase cure rates.

Hybrid Enzymes Containing Dihydrofolate Reductase Fused to Other Drug-Resistant Proteins

Although it is possible to express two or even possibly three cDNAs contained in a retroviral vector, using IRES sequences that allow translation of these messages, often the expression of one of these proteins is not optimal [see Sadelain et al., 44; this volume]. A new development based on the knowledge that DHFR gene exists in Plasmodia fused to the thymidylate synthase gene [45], has been the construction of retroviral vectors containing fusion cDNAs, separated by a small spacer oligonucleotide. An L22Y DHFR cDNA fused to wt TS cDNA has been reported [46], and recently we reported on a F22L/F31S cDNA fused to a human cytidine deaminase cDNA retroviral construct [47]. This construct provided resistance to both MTX and cytosine arabinoside in 3T3 cells and in mouse and human hematopoietic progenitors, as measured by survival of CFU-GM colonies. A detailed analysis of the steady-state kinetics of the recombinant fusion protein compared to recombinant DHFR and cytidine deaminase enzymes showed that the fusion protein had properties (K_m, V_{max}) similar to the individual proteins; the K_i for MTX was also unchanged, although the fusion protein was a little less sensitive to TMTX than the recombinant F22/F31S enzyme [47]. Recent unpublished studies from our laboratory show that mice are protected from the combined toxicity of MTX and cytosine arabinoside if marrow is transfected with the cDNA containing the hybrid cDNA.

Proof of Principle: Retroviral Transduction of Marrow Cells Allows High-Dose Chemotherapy Posttransplantation, Leading to an Improved Cure Rate

Although in principle, the safe use of chemotherapy posttransplantation after high-dose chemotherapy should lead to improved survival of patients, it was important to demonstrate this in a model system [48]. Mice bearing the transplanted E0771 breast tumor were treated with lethal doses of cyclophosphamide and rescued from toxicity by administration of bone marrow transduced with the F31S retroviral construct. Animals receiving marrow not transduced with mutant DHFR cDNA died from MTX toxicity, whereas mice transduced with mutant DHFR cDNA containing marrow were able to tolerate MTX treatment posttransplantation; 44% of the mice had no demonstrable tumor when sacrificed on day 52. Another control group of mice treated with cyclophosphamide and transplanted but not treated with MTX posttransplantation succumbed to tumor regrowth. These data provide a strong rationale for the use of drug resistance genes to protect marrow from drug toxicity because the increase in dose tolerated may result in an improved cure rate of chemosensitive tumors.

Are We Ready for Clinical Trials Using Marrow Dihydrofolate Reductases

As pointed out in this review as well as in other chapters in the volume, recent developments have now encouraged the initiation of clinical trials, using variants of DHFR for hematopoetic progenitor cell transduction to protect patients from MTX and TMTX toxicity: (1) variant DHFR cDNAs have been generated that produce a high level of resistance to these antifolates; (2) retroviral vectors have been developed that appear to be safe from recombination events, and express the cDNAs at high levels in cells after transduction; (3) transfection efficiencies of human CD34+ cells have increased, using higher titers of virus, repeated exposures to viral supernatants, and fibronectin fragment-coated vessels; (4) and the use of coexpressed green fluorescent protein or other proteins that allow selection in vitro, may be used to preselect cells that have been successfully transduced in vitro. Major obstacles still to be overcome are loss of expression of DHFR cDNA in vivo, and defective homing of transduced cells as a consequence of prolonged incubation in vitro [49]. The possibility of obtaining prolonged expression in patients after transduction with the use of continued MTX or TMTX-selective pressure may overcome some of these obstacles. Lessons from initial trials in patients transduced with MDR cDNA emphasize the need to include a control group (nontransduced) to compare tolerance of drug treatment; surprisingly, even nonsuccessfully transduced patients tolerated reasonable doses of paclitaxel in this trial [50]. Additionally, it would be useful to extend mouse studies to dogs, with spontaneous tumors (breast, lymphoma osteosarcoma), to obtain further proof of principle, as patients with these tumor types would be targeted for initial gene therapy approaches in humans.

References

1. Bertino JR: Karnofsky Memorial Lecture: Ode to methotrexate. J Clin Oncol 1993;11:5–14.
2. Bertino JR, Kamen B, Romanini A: Folate antagonists; in Holland JR, Bast RC Jr, Motton DL, Frei E II, Kufe DW, Weichselbaum RR (eds): Cancer Medicine, ed 4. Philadelphia, Williams and Wilkens, 1997, pp 907–922.
3. Hoffmeister RT: Methotrexate treatment of rheumatoid arthritis, 15 years experience. Am J Med 1983;30:69–73.
4. McDonald CJ: Hyperproliferative skin diseases; in McDonald CJ (ed): Immunomodulatory and Cytotoxic Agents in Dermatology. New York, Dekker, 1997, pp 267–282.
5. Blakley RL: Eukaryotic dihydrofolate reductase. Adv Enzymol Relat Areas Mol Biol 1995;70: 23–102.
6. Patel M, Sleep SI, Lewis WS, Spencer HT, Mareya SM, Sorrentino BP, Blakley RL: Comparison of the protection of cells from antifolates by transduced human dihydrofolate reductase mutants. Hum Gene Ther 1997;8:2069–2077.
7. Gorlick R, Goker E, Trippett T, Waltham M, Banerjee D, Bertino JR: Drug therapy: Intrinsic and acquired resistance to methotrexate in acute leukemia. New Engl J Med 1996;335:1041–1048.
8. Spencer HT, Sorrentino BP, Pui C-H, Chunduru SK, Sleep SEH, Blakley RL: Mutations in the gene for human dihydrofolate reductase: An unlikely cause of clinical relapse in pediatric leukemia after therapy with methotrexate. Leukemia 1996;10:439–446.
9. Fanin R, Banerjee D, Volkenandt M, Waltham M, Li WW, Dicker AP, Schweitzer BI, Bertino JR: Mutations leading to antifolate resistance in Chinese hamster ovary cells after exposure to the alkylating agent ethylmethanesulfonate. Molec Pharmacol 1993;44:13–21.
10. Oefner C, D'Arcy A, Winkler FK: Crystal structure of human dihydrofolate reductase complexed with folate. Eur J Biochem 1988;124:377–385.
11. Haber DA, Beverley SM, Kiely ML, Schimke RT: Properties of an altered dihydrofolate reductase encoded by amplified genes in cultured mouse fibroblasts. J Biol Chem 1981;256:9501–9510.
12. Simonsen CC, Levinson AD: Isolation and expression of an altered mouse dihydrofolate reductase cDNA. Proc Natl Acad Sci USA 1983;80:2495–2499.
13. Srimatkandada S, Schweitzer BI, Moroson BA, Dube S, Bertino JR: Amplification of a polymorphic dihydrofolate reductase gene expressing an enzyme with decreased binding to methotrexate in a human colon carcinoma cell line, HCT-8R4, resistant to this drug. J Biol Chem 1989;264:3524–3528.
14. Goldie JH, Krystal G, Hartley D, Gudauskas G, Dedhar S: A methotrexate insensitive variant of folate reductase present in two lines of methotrexate-resistant L5178Y cells. Eur J Cancer 1980;16: 1539–1546.
15. McIvor RS, Simonsen CC: Isolation and characterization of a variant dihydrofolate reductase cDNA from methotrexate-resistant murine L5178Y cells. Nucleic Acids Res 1990;18:7025–7032.
16. Melera PW, Davide JP, Oen H: Antifolate-resistant Chinese hamster cells. Molecular basis for the biochemical and structural heterogeneity among dihydrofolate reductases produced by drug-sensitive and drug-resistant cell lines. J Biol Chem 1988;263:1978–1990.
17. Dicker AP, Volkenandt M, Schweitzer BI, Banerjee D, Bertino JR: Identification and characterization of a mutation in the dihydrofolate reductase gene from the methotrexate-resistant Chinese hamster ovary cell line Pro-3 MTX^{RIII}. J Biol Chem 1990;265:8317–8321.
18. Dicker AP, Waltham MC, Volkenandt M, Schweitzer BI, Otter GM, Schmid FA, Sirotnak FM, Bertino JR: Methotrexate resistance in an in vivo mouse tumor due to a non-active-site dihydrofolate reductase mutation. Proc Natl Acad Sci USA 1993;90:11797–11801.
19. Ercikan-Abali EA, Waltham MC, Dicker AP, Schweitzer BI, Gritsman H, Banerjee D, Bertino JR: Variants of human dihydrofolate reductase with substitutions at leucine-22: Effect on catalytic and inhibitor binding properties. Mol Pharmacol 1996;49:430–437.
20. Morris JA, McIvor RS: Saturation mutagenesis at dihydrofolate reductase codons 22 and 31, a variety of amino acid substitutions conferring methotrexate resistance. Biochem Pharmacol 1994;47:1207.
21. Nakano T, Spencer HT, Appleman JR, Blakely RL: Critical role of phenylalanine 34 of human dihydrofolate reductase in substrate and inhibitor binding and in catalysis. Biochemistry 1994;33: 9945–9952.

22 Ercikan-Abali EA, Mineishi S, Yong T, Nakahara S, Waltham MC, Banerjee D, Chen W, Sadelain M, Bertino JR: Active site-directed double mutants of dihydrofolate reductase. Cancer Res 1996; 56:4142–4145.
23 Tsay J-T, Appleman JR, Beard WA, Prendergast NJ, Delcamp TJ, Freisheim JH, Blakley RL: Kinetic investigation of the functional role of phenylalanine-31 of recombinant human dihydrofolate reductase. Biochemistry 1990;29:6428–6436.
24 Thompson PD, Freisheim JH: Conversion of arginine to lysine at position 70 of human dihydrofolate reductase: Generation of a methotrexate-insensitive mutant enzyme. Biochemistry 1991;30:8124–8130.
25 Tan X, Huang S, Ratnam M, Thompson PD, Freisheim JH: The importance of loop region residues 40–46 in human dihydrofolate reductase as revealed by site-directed mutagenesis. J Biol Chem 1990; 265:8027–8032.
26 Bullerjahu AME, Freisheim JH: Site directed deletion mutants of a carboxy-terminal region of human dihydrofolate reductase. J Biol Chem 1992;267:864–870.
27 Huang S, Delcamp TJ, Tan X, Smith PL, Prendergast NJ, Freisheim JH: Effects of conversion of an invariant tryptophan residue to phenylalanine on the function of human dihydrofolate reductase. Biochemistry 1989;28:471–478.
28 Lewis WR, Cody V, Galtisky N, Luft JR, Pagborn W, Chunduru SK, Spencer HT, Appleman JR, Blakley RL: Methotrexate-resistant variants of human dihydrofolate reductase with substitutions of leucine 22. Kinetics, crystallography, and potential as selectable markers. J Biol Chem 1995;270: 5057–5064.
29 Chunduru SK, Cody V, Luft JR. Pangborn W, Appleman JR, Blakely RL: Methotrexate resistant variants of human dihydrofolate reductase. Effects of Phe^{30} substitutions. J Biol Chem 1994;269: 9547–9555.
30 Tslasty TD: Normal diploid human and rodent cells lack a detectable frequency of gene amplification. Proc Natl Acad Sci USA 1990;87:3132–3136.
31 Wright JA, Smith HS, Watt FM, Hancock MC, Hudson DL, Stark GR: DNA amplification is rare in normal human cells. Proc Natl Acad Sci USA 1990;87:1791–1795.
32 Allay JA, Spencer HT, Wildman SL, Beet JA, Blakley RL, Sorrentino BP: Sensitization of hematopoietic stem and progenitor cells to trimetrexate using nucleoside transport inhibitors. Blood 1997;90: 3546–3554.
33 Cline MJ, Stang H, Mercola K, Morse L, Ruprecht R, Browne J, Salse W: Gene transfer in intact animals. Nature 1980;284:422–425.
34 Carr F, Medina WD, Dube S, Bertino JR: Genetic transformation of murine bone marrow cells to methotrexate resistance. Blood 1983;62:180–185.
35 Williams DA, Hsieh K, DeSilva AS, Mulligan RC: Protection of bone marrow transplant recipients from lethal doses of methotrexate by the generation of methotrexate resistant bone marrow. J Exp Med 1987;166:210–218.
36 Corey CA, DeSilva AD, Holland CA, Williams DA: Serial transplantation of methotrexate resistant bone marrow: Protection of murine recipients from drug toxicity by progeny of transduced stem cells. Blood 1990;76:337–343.
37 Banerjee D, Schweitzer BI, Li MX, Volkenandt M, Waltham MW, Mineishi S, Zhao SC, Bertino JR: Transfection with cDNA encoding a Ser^{31} or Ser^{14} mutant human dihydrofolate reductase into Chinese hamster ovary and mouse marrow progenitor cells confers methotrexate resistance. Gene 1994;139:269–274.
38 Hantzopalos PA, Sullenger BA, Ungers G, Gilboa E: Improved gene expression upon transfer of the adenosine deaminase minigene outside the transcriptional unit of a retroviral vector. Proc Natl Acad Sci USA 1989;86:3519–3523.
39 Li MX, Hantzopoulos PA, Banerjee D, Zhao SC, Schweitzer BI, Gilboa E, Bertino JR: Comparison of the expression of a mutant dihydrofolate reductase under control of different internal promoters in retroviral vectors. Hum Gene Ther 1992;3:381–390.
40 Zhao SC, Banerjee D, Li MX, Schweitzer BI, Gilboa E, Bertino JR: Efficient transfer and persistent expression of mutant DHFR gene in normal mouse hematopoietic cells. Proc Am Assoc Cancer Res 1992;33:2955a.

41 Zhao SC, Li MX, Banerjee D, Schweitzer BI, Mineishi S, Gilboa E, Bertino JR: Long-term protection of recipient mice from lethal doses of methotrexate by marrow infected with a double copy vector retrovirus containing a mutant dihydrofolate reductase. Cancer Gene Ther 1994;1:27–33.

42 Li MX, Banerjee D, Zhao SC, Schweitzer BI, Mineishi S, Gilboa E, Bertino JR: Development of a retroviral construct containing a human mutated dihydrofolate reductase cDNA for hematopoietic stem cell transduction. Blood 1994;83:3403–3408.

43 Rao SS, Peters SO, Crittenden RB, Stewart FM, Ramshan HS, Quesenberry PJ: Stem cell transplantation in the normal nonmyeloablated host: Relationship between cell dose, schedule, and engraftment. Exp Hematol 1997;25:114.

44 Sadelain M, May C, Rivella S, Glade Bender J: Basic principles of gene transfer in hematopoietic stem cells; in Bertino JR (ed): Marrow Protection. Prog Exp Tum Res. Basel, Karger, 1999, vol 36, pp 1–19.

45 Fantz CR, Shaw D, Moore JG, Spencer HT: Retroviral coexpression of thymidylate synthase and dihydrofolate reductase confers fluoropyrimidine and antifolate resistance. Biochem Biophys Res Commun 1998;243:6–12.

46 Ivanetich KM, Santi DV: Bifunctional thymidylate synthase-dihydrofolate reductase in protozoa. FASEB J 1990;4:1591–1597.

47 Sauerbrey A, McPherson JP, Banerjee D, Bertino JR: Simultaneous resistance to methotrexate (MTX) and cytarabine conferred by expression of a double mutant dihydrofolate reductase – Cytidine deaminase fusion gene. Blood 1998;467a.

48 Zhao SC, Banerjee D, Mineishi S, Bertino JR: Post-transplant methotrexate administration leads to improved curability of mice bearing a mammary tumor transplanted with marrow transduced with a mutant dihydrofolate reductase cDNA. Hum Gene Ther 1997;8:903–909.

49 Gan OI, Murdoch B, Larochelli A, Dick JE: Differential maintenance of primitive human SCID-repopulating cells, clonogenic progenitors, and long term culture-initiating cells after incubation in human bone marrow stromal cells. Blood 1997;90:641–650.

50 Rahman Z, Kavanaugh J, Champlin R, et al: Chemotherapy immediately following autologous stem cell transplantation in patients with breast cancer. Clin Cancer Res 1998;4:2717–2722.

Joseph R. Bertino, Molecular Pharmacology and Therapeutics Program,
Sloan Kettering Institute for Cancer Research, 1275 York Avenue, New York, NY 10021 (USA)
Tel. +1 212 639 8230, Fax +1 212 639 2767, E-Mail bertinoj@mskcc.org

Augmentation of Methotrexate Resistance with Coexpression of Metabolically Related Genes

Shin Mineishi

Bone Marrow Transplant Division, Department of Medicine, University of Wisconsin, Madison, Wisc., USA

Introduction

Gene therapy offers exciting opportunities for improving treatment of many diseases. Investigators have made attempts to apply gene therapy for cancer treatment [1]. Protecting normal organs from toxicity of chemotherapeutic agents by gene transfer of drug resistance genes is a promising method of improving cancer therapy [2]. Bone marrow has been the primary target for drug resistance gene transfer because bone marrow toxicity is dose-limiting for many chemotherapeutic agents [3]. Besides, bone marrow is an organ which is relatively easy to harvest and easy to transplant. The observation made by Hryniuk et al. [4] that chemotherapy efficacy in certain malignancies is directly correlated with dose intensity, stimulated dose-intensive chemotherapy regimens with bone marrow protection, using either autologous bone marrow or peripheral stem cells for rescue with hematopoietic growth factor support. Successful gene transfer of drug resistance genes into hematopoietic progenitor cells may further protect bone marrow from toxicity of chemotherapy, and may make further dose increases possible, and thus improve treatment outcome.

In vivo selection of a nonselectable gene is another goal of drug resistance gene transfer into hematopoietic cells. The efficiency of gene transfer into hematopoietic cells is still insufficient for clinically meaningful gene therapy [5, 6]. If transduced clones can be enriched using in vivo drug selection, the problem of low transduction efficiency may be overcome, or at least compensated. By expressing a nonselectable gene together with a drug-selectable gene, in vivo enrichment of the second gene may be possible by selecting with the drug. This second gene can be a therapeutic gene, such as the glucocerebrosidase gene for Gaucher's disease, or the iduronidase gene for Hurler's syndrome. We and others have shown that in vivo drug selection is possible in mice [7, 8]. In our experiments, an altered dihydrofolate reductase (DHFR) gene (Ser31), which is a target enzyme of methotrexate (MTX), was used as an in vivo selectable marker. The enrichment of neomycin resistance gene (Neo-r) containing clones with MTX selection was demonstrated [7]. In vivo drug selection has yet to be demonstrated in large animals or humans.

In general, the degree of drug resistance is high when fibroblasts are used as target cells, but when hematopoietic cells are used, the large selective advantage is often lost. Thus, a larger margin of selectability, in other words, a high level of resistance, is always desirable. Continuing efforts have been made to improve the drug resistance conferred by transfer of a single gene, with some successes. Many different mutants of MDR1 (multidrug resistance), DHFR, and other drug resistance genes have been made and resistance conferred by them have been compared [9–14].

Vectors for Gene Transfer

In order to achieve high selectability with drug resistance genes, a method for efficient transduction is needed. For bone marrow transduction with a foreign gene, retrovirus-mediated gene transfer is currently the method of choice. Although retrovirus vectors cannot transduce nondividing cells, in dividing cells retrovirally transduced genes are integrated into the genome of the target cells and thus the transgenes are relatively stable [15]. Because retrovirus vectors can accommodate relatively large genes – up to 8 kb or so – they are suited for coexpression of metabolically related genes. Also, if more than one gene is expressed by a vector, it is generally accepted that using an internal ribosome entry site and expressing the mRNA by the viral long terminal repeat promoter is more desirable than using internal promoters, as the latter method creates more than one transcriptional unit and there may be interference between them [16]. Other viral vectors, including adenovirus and adeno-associated virus (AAV), are under active investigation. Results with AAV seem promising [17–20]. Wild-type AAV, not recombinant AAV, is known to be integrated into the genome in a site-specific manner [17]. AAV is not known to be associated with any mammalian cancer, which is an advantage of this vector. It seems that it requires the target cells to be in cycle for transduction. A disadvantage of this vector is that it can accommodate only up to 5-kb gene substitutions. Adenovirus, on the other hand, does not infect bone marrow cells efficiently and high multiplicities of infection are required for marrow cell transduction. Adenoviral vectors also have some drawbacks, including a relatively short duration of expression of the transferred gene, antigenicity which may preclude repeated administration, and toxicity associated with high doses of virus [21, 22]. In order to transduce early hematopoietic progenitor cells which rarely divide, vectors which can transduce nondividing cells, such as HIV-based vectors [23], are under study.

Early Hematopoietic Progenitor Cells as Targets of Gene Transfer

Early hematopoietic progenitor cells have been characterized in more detail recently [24–26]. As stated above, early hematopoietic progenitor cells (stem cells) rarely divide. To stimulate cell cycle in early progenitor cells in order to increase transduction efficiency, different combinations of growth factors have been used, including kit ligand, IL-1, IL-3, and IL-6 [27–31]. Recently, flt-3 ligand has been shown to increase engraftment potential after transduction [32–34]. To improve transduction efficiency, stromal cell layers have been used [35]. Fibronectin fragment CH-296 was used to facilitate transduction into hematopoietic progenitor cells [36, 37]. Peripheral blood stem cells and cord blood stem cells also have been used for gene therapy as alternative sources of hematopoietic progenitor cells. They may be better targets for retrovirus-mediated gene transfer than bone marrow progenitor cells [38–40]. With a combination of these methods and refinement of procedures, we are now successful in transducing 40–60% of human CD34+CD38− cells consistently [Mineishi, S., unpubl. data].

DHFR Gene and Its Mutants

Theoretically, any gene which expresses a protein that confers resistance to chemotherapeutic agents is a candidate for drug resistance gene transfer [1]. For clinical use, however, the gene should have the following attributes: (1) the gene must be well characterized; (2) bone marrow toxicity has to be one of or the dose-limiting toxicity of the drug; (3) the drug to which the bone marrow cells become resistant with transfer of the gene has to have activity against the cancer under treatment. The genes studied most extensively in experimental gene transfer systems are DHFR and the MDR1 [2, 7–9].

DHFR has several advantages over MDR1. (1) DHFR is a small gene, only 564 bp, as opposed to 4 kb of MDR1 gene. It is easily accommodated in viral vectors. (2) DHFR mutants are available that result in a high level of resistance [9]. (3) MTX is a good candidate for an in vivo selection drug, as it is relatively nontoxic at low doses, and is commonly used in non-malignant disease settings (e.g. psoriasis, rheumatoid arthritis) [41]. (4) DHFR is related to many metabolic pathways [42] and thus MTX resistance can be augmented with coexpression of metabolically related genes.

Overexpression of wild-type DHFR or mutations in DHFR, which leads to decreased binding of MTX, results in resistance to this drug. Of these mutants, Arg22, Phe22(Leu→Arg or Phe at 22, respectively), or Ser31(Phe→Ser at 31) have been tested most extensively as dominant selectable markers. The Arg22 mutant binds to MTX very poorly, and thus imparts a very high level of resistance to the drug but is catalytically a poor enzyme. The Ser31 mutant, which was characterized by us, confers as a high level of resistance to MTX as Arg22, and has good catalytic properties [43]. We generated other mutant forms of DHFR and among them, a double mutant, F22S31(Leu→Phe at 22, Phe→Ser at 31), was found to confer the highest level of resistance to MTX in transduced cells [9].

In 1987, Williams et al. [44] reported successful gene transfer with the Arg22 DHFR into mouse bone marrow using a retroviral vector. The transduced bone marrow cells were transplanted into lethally irradiated mice and successful protection of bone marrow cells was demonstrated with MTX challenge. Transduction of a primitive hematopoietic stem cell was shown by the fact that animals receiving secondary bone marrow transplants from transduced mice were also protected from MTX toxicity [45].

Recently, we completed a study using the Ser31 mutant DHFR. Mice transplanted with bone marrow transduced with the DC/SVhDHFR S31 vector, which contains the Ser 31 mutant, and treated with MTX had a survival advantage, associated with decreased hematologic toxicity, over control mice which received mock-transduced bone marrow [7]. Infection with this construct of human CD34-selected peripheral stem cells as well as cord blood cells also resulted in MTX resistance [21].

More recently, we established a tumor bearing mouse model to test the therapeutic efficacy of bone marrow protection by drug resistance gene transfer. In this model, we implanted mice with a mammary tumor cell line. Mice received either mock-transduced or drug-resistance-gene-transduced bone marrow cells and subsequently received MTX after bone marrow transplantation. In this model, we demonstrated that additional therapeutic advantage can be obtained by protection of bone marrow with drug resistance gene transfer [46].

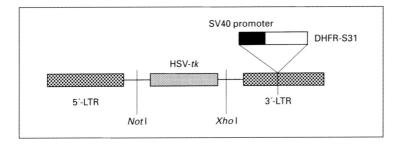

Fig. 1. Structure of the DC/SV6S31*tk* retrovirus vector. DC/SV6S31 and DC/SV6S31*gpt* have the same structure except that HSV*tk* is replaced with Neo-r or XGPRT, respectively. Reproduced with the permission of Stockton Press from [47].

Coexpression of Metabolically Related Genes

Coexpression of a metabolically related gene is another possible way to augment resistance to chemotherapy drugs imparted by a drug resistance gene transfer [47].

A Mutated DHFR cDNA and HSVtk cDNA

MTX inhibits DHFR and depletes TTP pools in the cells. Also, as polyglutamated forms, it inhibits purine biosynthesis. Thus, MTX has effects on both purine and pyrimidine synthesis pathways [47]. It can be expected that if one or both of these pathways are rescued, toxicity of MTX may be alleviated. It has been known that thymidine kinase can be selected with HAT medium, which contains hypoxanthine, aminopterin, and thymidine [48]. Aminopterin is very similar to MTX in potency and in its chemical structure [2]. Recently, Sorrentino's group [49] at St. Jude showed that adding thymidine transport inhibitors augments the selectability of an MTX resistance gene in the mouse model, providing evidence that thymidine salvage may alleviate the toxicity of MTX. We constructed a retroviral vector which expressed a mutated DHFR, Ser31, together with the herpes simplex virus thymidine kinase (HSV*tk*) gene, DC/SV6S31*tk* (fig. 1) [47]. The HSV*tk* gene is a suicide gene which activates a prodrug ganciclovir (GCV). Also, because it converts thymidine into dTMP which is subsequently converted into dTTP, it salvages thymidine effectively, which is depleted as a result of MTX treatment. Using this vector, we achieved 10-fold higher resistance to MTX in infected NIH3T3 cells, compared to our conventional MTX resistance vector, DC/SV6S31, which carries an Ser31

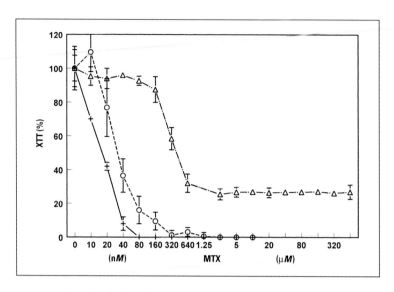

Fig. 2. XTT colorimetric assay of the cytotoxicity of MTX. 3T3 cells (+), 3T3 cells infected with the DC/SV6S31 vector (○), and the DC/SV6S31*tk* vector (△). XTT, sodium 3'-(1-((phenylamino)-carbonyl)-3,4-tetrazolium)-bis(4-methoxy-6-nitro)benzene-sulfonic acid hydrate is a reagent for colorimetric assay. Reproduced with the permission of Stockton Press from [47].

mutated DHFR cDNA and Neo-r instead of HSV*tk*. This augmented resistance was lost when dialyzed fetal bovine serum (FBS), which is depleted of thymidine, was used instead of regular FBS. Interestingly, about 20% of the cells transduced with this vector, survived with MTX concentrations of 200 μM or more (fig. 2). The DC/SV6S31*tk* vector was also shown to confer additional protection to MTX to transduced mouse bone marrow cells compared to the DC/SV6S31 vector, as demonstrated by increased numbers of MTX-resistant CFU-GM colonies.

One of the concerns associated with bone marrow transduction with drug resistance genes is that tumor cells present in peripheral blood or in the bone marrow may also be transduced with the drug resistance gene, thus producing drug-resistant tumor cells. To avoid this problem, extreme precaution needs to be taken to exclude the possibility that preparations of bone marrow/peripheral stem cells are contaminated with tumor cells. Use of vector constructs which contain suicide genes, such as HSV*tk*, has been proposed as a method to eradicate the transduced tumor cells selectively, if necessary, as a safeguard [50]. The vector we constructed, DC/SV6S31*tk*, contains the HSV*tk* gene, which serves for this purpose (fig. 3). As a consequence, this vector is

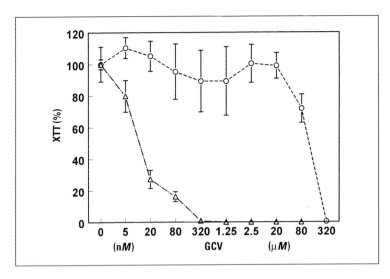

Fig. 3. XTT assay of the cytotoxicity of GCV. 3T3 cells infected with the DC/SV6S31 vector (○) and the DC/SV6S31*tk* vector (△). Reproduced with the permission of Stockton Press from [47].

bifunctional, i.e. it can be selected either negatively or positively in vivo by using MTX or GCV. For example, this vector may be used to transduce effector cells, such as cytotoxic T cells, to convey biological effects, such as anti-tumor immunity. To enrich the functional clone, MTX in vivo selection can be used. These effector cells can be eradicated by using GCV if they exhibit side effects [51].

A Mutated DHFR and XGPRT

We have also constructed another vector DC/SV6S31*gpt*, which contains xanthine guanine phosphoribosyl transferase (XGPRT) cDNA and a mutated DHFR cDNA, Ser31 [52]. XGPRT is a bacterial enzyme which is similar to mammalian hypoxanthine guanine phosphoribosyl transferase (HGPRT), one of the enzymes of the purine salvage pathway. As stated above, MTX also inhibits purine synthesis. Thus, because this gene salvages purines, MTX toxicity is partially relieved. Coexpression of XGPRT augmented resistance 4- to 8-fold measured by the increase of ID_{50} to MTX in transduced cells compared to the DC/SV6S31 vector (fig. 4). Again, the advantage of the augmented resistance is lost when dialyzed serum is used instead of regular FBS [52].

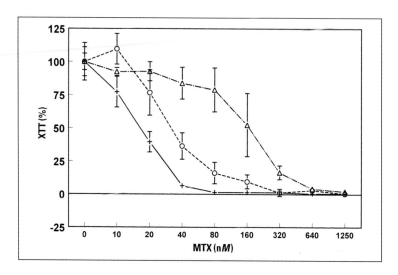

Fig. 4. XTT assay of the cytotoxicity of MTX. 3T3 cells (+), 3T3 cells infected with the DC/SV6S31 vector (○), and the DC/SV6S31*gpt* vector (△). Reproduced with the permission of Appleton & Lange from [52].

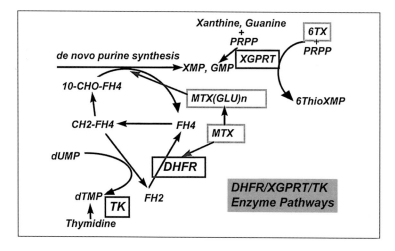

Fig. 5. Enzymatic pathways related to DHFR, HSV*tk*, and XGPRT. Reproduced with the permission of Appleton & Lange from [52].

It has been observed that retroviral packaging cell lines, such as GP-AM12 or GP-E86, which contain XGPRT gene as a dominant selectable marker [53, 54], have very high resistance to MTX, and it is virtually impossible to select the transduced cells with a mutant DHFR as a selectable marker when the target cells are these cell lines [52]. This probably is related to very high copy numbers of XGPRT gene in these cell lines, and provides additional evidence that coexpression of XGPRT rescues MTX toxicity. XGPRT also may be used as a suicide gene with 6-thioxanthine (6-TX), as XGPRT, unlike HGPRT, utilizes a nontoxic purine, 6-TX, as a substrate, and converts it to a toxic substance [55]. However, we could not demonstrate a suicide function with our construct. A schematic diagram of these enzyme pathways is shown in figure 5.

Conclusions

Expression of a gene which is metabolically related to a drug resistance gene is one possible way of augmenting drug resistance which is conferred by the drug resistance gene. The MTX-DHFR system is well suited for this purpose, as it is related to many enzymatic pathways and relatively easy to modify. By coexpression of HSV*tk* or XGPRT together with the Ser31 DHFR mutant, up to 10-fold augmented resistance was achieved. Combinations of the genes other than presented here are currently under active investigation.

References

1 Anderson WF: Gene therapy for cancer. Hum Gene Therapy 1995;5:1–2.
2 Bertino JR: Ode to methotrexate. J Clin Oncol 1993;11:5–14.
3 Banerjee D, Zhao SC, Li M-X, Schweitzer BI, Mineishi S, Bertino JR: Gene therapy utilizing drug resistance genes: A review. Stem Cells 1994;12:378–385.
4 Hryniuk W, Bush H: The importance of dose intensity in chemotherapy of metastatic breast cancer. J Clin Oncol 1984;2:1281–1288.
5 Brenner MK, Rill DR, Holladay MS, Heslop HE, Moen RC, Buschle M, Krance RA, Santana VM, Anderson WF, Ihle JN: Gene marking to determine whether autologous marrow infusion restores long-term hematopoiesis in cancer patients. Lancet 1993;342:1134–1137.
6 Dunbar CE, Cottler-Fox M, O'Shaughnessy JA, Doren S, Carter C, Berenson R, Brown S, Moen RC, Greenblatt J, Stewart FM, Leitman SF, Wilson WH, Cowan K, Young NS, Nienhuis AW: Retrovirally marked CD34-enriched peripheral blood and bone marrow cells contribute to long-term engraftment after autologous transplantation. Blood 1995;85:3048–3057.
7 Li M-X, Banerjee D, Zhao SC. Schweitzer BI, Mineishi S, Gilboa E, Bertino JR: Development of a retroviral construct containing a human mutated dihydrofolate reductase cDNA for hematopoietic stem cell transduction. Blood 1994;83:3403–3408.
8 Sorrentino BP, Brandt SJ, Bodine D, Gottesman M, Pastan I, Cline A, Nienhuis AW: Selection of drug-resistant bone marrow cells in vivo after retroviral transfer of human Mdr1. Science 1992;257:99–103.
9 Ercikan-Abali E, Mineishi S, Tong Y, Nakahara S, Waltham MC, Banerjee D, Chen W, Sadelain M, Bertino JR: Active site-directed double mutants of dihydrofolate reductase. Cancer Res 1996;56:4142–4145.
10 Gros P, Dhir R, Croop J, Talbot F: A single amino acid substitution strongly modulates the activity and substrate specificity of the mouse mdr1 and mdr3 drug efflux pumps. Proc Natl Acad Sci USA 1991;88:7289–7293.

11 Spencer HT, Sleep SE, Rehg JE, Blakley RL, Sorrentino BP: A gene transfer strategy for making bone marrow cells resistant to trimetrexate. Blood 1996;87:2579–2587.
12 Schweitzer BI, Srimatkandada S, Gritsman H, Sheridan R, Venkataraghavan R, Bertino JR: Probing the role of two hydrophobic active site residues in the human dihydrofolate reductase by site-directed mutagenesis. J Biol Chem 1989;264:20786–20795.
13 Dicker AP, Waltham MC, Volkenandt M, Schweitzer BI, Otter GM, Schmid FA, Sirotnak FM, Bertino JR: Methotrexate resistance in an in vivo mouse tumor due to a non-active-site dihydrofolate reductase mutation. Proc Natl Acad Sci USA 1993;90:11797–11801.
14 Schweitzer BI, Dicker AP, Bertino JR: Dihydrofolate reductase as a therapeutic target. FASEB J 1990;4:2441–2452.
15 Crystal RG: Transfer of genes to humans: Early lessons and obstacles to success. Science 1995;270: 404–410.
16 Morgan RA, Couture L, Elroy-Stein O, Ragheb J, Moss B, Anderson WF: Retroviral vectors containing putative internal ribosome entry sites: Development of a polycistronic gene transfer system and applications to human gene therapy. Nucleic Acid Res 1992;20:1293–1299.
17 Kotin R, Sinscalco M, Samulski RJ, Zhu X, Hunter L, Laughlin C, McLaughlin S, Muzyczka N, Rocchi M, Berns K: Site-specific integration by adeno-associated virus. Proc Natl Acad Sci USA 1990;87:2211–2215.
18 Zhou SZ, Broxmeyer H, Cooper S, Harrinton MA, Srivasta A: Adeno-associated virus 2-mediated gene transfer in murine hematopoietic progenitor cells. Exp Hematol 1993;21:928–933.
19 Chatterjee S, Johnson PR, Wong KK: Dual-target inhibition of HIV-1 in vitro by means of adeno-associated virus antisense vector. Science 1992;258:1485–1488.
20 Ponnazhagan S, Yoder MC, Srivastava A: Adeno-associated virus type 2-mediated transduction of murine hematopoietic cells with long-term repopulating ability and sustained expression of a human globin gene in vivo. J Virol 1997;3098–3104.
21 Rosenfeld MA, Yoshimura K, Trapnell BC, Yoneyama K, Rosenthal ER, Dalemans W, Fukayama M, Bargon J, Stier LE, Stratford-Perricaudet L, Perricaudet M, Guggino WB, Pavirani A, Lecocq J-P, Crystal RG: In vivo transfer of the human cystic fibrosis transmembrane conductance regulator gene to the airway epithelium. Cell 1992;68:143–155.
22 Li Q, Kay M, Finegold M, Stratford-Perricaudet L, Woo S: Assessment of recombinant adenoviral vectors for hepatic gene therapy. Hum Gene Ther 1993;4:403–409.
23 Naldini L, Blomer U, Gallay P, Ory D, Mulligan R, Gage FH, Verma IM, Trono D: In vivo gene delivery and stable transduction of nondividing cells by a lentiviral vector. Science 1996;272:263–267.
24 Moore MAS: Clinical implications of positive and negative hematopoietic stem cell regulators. Blood 1990;78:1–19.
25 Larochelle A, Vormoor J, Hanenberg H, Wang JCY, Bhatia M, Lapidot T, Moritz T, Murdoch B, Xiao XL, Kato I, Williams DA, Dick JE: Identification of primitive human hematopoietic cells capable of repopulating NOD/SCID mouse bone marrow: Implications for gene therapy. Nat Med 1996;2:1329–1337.
26 Sitnicka E, Ruscetti FW, Priestley GV, Wolf NS, Bartelmez SH: Transforming growth factor β1 directly and reversibly inhibits the initial cell division of long-term repopulating hematopoietic stem cells. Blood 1996;88:82–88.
27 Flasshove M, Banerjee D, Mineishi S, Schlafstein M, Bertino JR Moore MAS: Retrovirally mediated gene transfer of a mutant human dihydrofolate reductase (DHFR) gene into progenitors from human peripheral blood (PB). Blood 1995;85:566–574.
28 Kuga T, Sakamaki S, Matsunaga T, Hirayama Y, Kuroda H, Takahashi Y, Kusakabe T, Kato I, Nittsu Y: Fibronectin fragment-facilitated retroviral transfer of the glutathione-S-transferase gene into CD34+ cells to protect them against alkylating agents. Hum Gene Ther 1997;8:1901–1910.
29 Glimm H, Kiem H-P, Darovsky B, Storb R, Wolf J, Diehl V, Mertelsmann R, von Kelle C: Efficient gene transfer in primitive CD34+CD38lo human bone marrow cells reselected after long-term exposure to GaLV-pseudotyped retroviral vector. Hum Gene Ther 1997;8:2079–2086.
30 Szilvassy SJ, Fraser CC, Eaves CJ, Lansdorp PM, Eaves C, Humphries RK: Retrovirus-mediated gene transfer to purified hematopoietic stem cells with long-term lympho-myelopoietic repopulating ability. Proc Natl Acad Sci USA 1989;86:8798–8802.

31 Bodine DM, Karlsson S, Nienhuis AW: Combination of interleukins 3 and 6 preserves stem cell function in culture and enhance retrovirus-mediated gene transfer into hematopoietic stem cells. Proc Natl Acad Sci USA 1989;86:8897–8901.

32 Haylock DN, Horsfall MJ, Dowse TL, Ramshaw HS, Niutta S, Protopsaltis S, Peng L, Burrell C, Rappold I, Buhring HJ, Simmons PJ: Increased recruitment of hematopoietic progenitor cells underlies the ex-vivo expansion potential of Flt3 ligand. Blood 1997;90:2260–2272.

33 Yoneyama Y, Ku H, Lyman SD, Ogawa M: In vitro expansion of hematopoietic progenitors and maintenance of stem cells: Comparison between flt3/flk-2 ligand and kit ligand. Blood 1997;89: 1915–1921.

34 Dao MA, Hannum CH, Kohn DB, Nolta JA: Flt3 ligand preserves the ability of human CD34+ progenitors to sustain long-term hematopoiesis in immune-deficient mice after ex-vivo retroviral-mediated transduction. Blood 1997;89:446–456.

35 Moore KA, Deisseroth AB, Reading CL, Williams DE, Belmont JW: Stromal support enhances cell-free retroviral vector transduction of human bone marrow long-term culture-initiating cells. Blood 1992;79:1393–1399.

36 Moritz T, Patel VP, Williams DA: Bone marrow extracellular matrix molecules improve gene transfer into human hematopoietic cells via retroviral vectors. J Clin Invest 1994;93:1451–1457.

37 Hanenberg H, Xiao LX, Diloo D, Hashino K, Kato I, Williams DA: Colocalization of retrovirus and target cells on specific fibronectin fragments increases genetic transduction of mammalian cells. Nat Med 1996;2:876–882.

38 Lu L, Xiao M, Shen R-N, Grigsby S, Broxmeyer HE: Enrichment, characterization, and responsiveness of single primitive CD34 human umbilical cord blood hematopoietic progenitors with high proliferative and replating potential. Blood 1993;81:41–48.

39 Moritz T, Keller DC, Williams DA: Human cord blood cells as targets for gene transfer: Potential use in genetic therapies of severe combined immunodeficiency disease. J Exp Med 1993;178:529–536.

40 Bregni M, Magni M, Siena S, Di Nicola M, Bonadonna G, Gianni AM: Human peripheral blood hematopoietic progenitors are optimal targets of retroviral-mediated gene transfer. Blood 1992;80: 1418–1422.

41 Mineishi S, Bertino JR: Methotrexate; in McDonald C (ed): Immunomodulatory and Cytotoxic Agents in Dermatology. New York, Marcel Dekker, 1996, pp 9–20.

42 Allegra CJ Antifolates; in Chabner BA, Collins JM (eds): Cancer Chemotherapy: Principle and Practice. Philadelphia, Lippincott, 1990, pp 110–153.

43 Banerjee D, Schweitzer BI, Li MX, Volkenandt M, Waltham MW, Mineishi S, Zhao SC, Bertino JR: Transfection with a cDNA encoding a Ser^{31} or Ser^{34} mutant human dihydrofolate reductase into Chinese hamster ovary and mouse marrow progenitor cells confers methotrexate resistance. Gene 1994;139:269–274.

44 Williams DA, Hsieh K, DeSilva A, Mulligan RC: Protection of bone marrow transplant recipients from lethal doses of methotrexate by the generation of methotrexate-resistant bone marrow. J Exp Med 1987;166:210–218.

45 Corey CA, DeSilva AD, Holland CA, Williams DA: Serial transplantaion of methotrexate-resistant bone marrow: Protection of murine recipients from drug toxicity by progeny of transduced stem cells. Blood 1990;75:337–343.

46 Zhao SC, Banerjee D, Mineishi S, Bertino JR: Post transplant methotrexate administration leads to improved curability of mice bearing a mammary tumor transplanted with marrow transduced with a mutant human dihydrofolate reductase cDNA. Hum Gene Ther 1997;8:903–909.

47 Mineishi S, Nakahara S, Takebe N, Banerjee D, Zhao SC, Bertino JR: Co-expression of herpes simplex virus thymidine kinase gene potentiates methotrexate resistance conferred by transfer of a mutated dihydrofolate reductase gene. Gene Ther 1997;4:570–576.

48 Littlefield JW: Selection of hybrids from matings of fibroblasts in vitro and their presumed recombinants. Science 1964;145:709–710.

49 Allay JA, Spencer HT, Wilkinson SL, Belt JA, Blakley RL, Sorrentino BP: Sensitization of hematopoietic stem and progenitor cells to trimetrexate using nucleoside transport inhibitors. Blood 1997; 90:3546–3554.

50   Sugimoto Y, Sato S, Tsukahara S, Suzuki M, Okochi E, Gottesman MM, Pasten I, Tsuruo T: Coexpression of multidrug resistance gene (Mdr1) and herpes simplex virus thymidine kinase gene in a bicistronic retroviral vector Ha-MDR-IRES-TK allows selective killing of Mdr1-transduced human tumors transplanted in nude mice. Cancer Gene Ther 1997;4:51–58.

51   Bonini C, Ferrari G, Verzeletti S, Servida P, Zappone E, Ruggieri L, Ponzoni M, Rossini S, Mavilio F, Traversari C, Bordignon C: HSV-TK gene transfer into donor lymphocytes for control of allogeneic graft-versus-leukemia. Science 1997;276:1719–1724.

52   Mineishi S, Nakahara S, Takebe N, Banerjee D, Zhao SC, Bertino JR: Purine biosynthesis pathway rescue by xanthine guanine phosphoribosyl transferase (xgprt) potentiates methotrexate resistance conferred by transfer of a mutated dihydrofolate reductase gene. Cancer Gene Ther 1998;5:144–149.

53   Markowitz D, Goff S, Bank A: A safe packaging line for gene transfer: Separating viral genes on two different plasmids. J Virol 1988;62:1120–1124.

54   Markowitz D, Goff S, Bank A: Construction and use of a safe and efficient amphotropic packaging cell line. Virology 1988;167:400–406.

55   Mroz PJ, Moolten FL: Retrovirally transduced *Escherichia coli* gpt genes combine selectability with chemosensitivity capable of mediating tumor eradication. Hum Gene Ther 1993;4:589–595.

Shin Mineishi, MD, K6/564 CSC, University of Wisconsin,
Highland Avenue, Madison, WI 53792 (USA)
Tel. +1 608 265 8688 or 8690, Fax +1 608 262 0759, E-Mail sxm@heme.medicine.wisc.edu

Protection of Bone Marrow Cells from Toxicity of Chemotherapeutic Agents Targeted toward Thymidylate Synthase by Transfer of Mutant Forms of Human Thymidylate Synthase cDNA

Debabrata Banerjee, Youzhi Tong, Xinyue Liu-Chen, Gina Capiaux, Emine A. Ercikan-Abali, Naoko Takebe, Owen A. O'Connor, Joseph R. Bertino

Program of Molecular Pharmacology and Experimental Therapeutics, Sloan Kettering Institute for Cancer Research, New York, N.Y., USA

Use of Mutant Human Thymidylate Synthase cDNA for Myeloprotection

Myelotoxicity or bone marrow suppression is the major dose-limiting toxicity of several clinically relevant chemotherapeutic agents [1, 2]. Use of recombinant hematopoietic growth factors or cytokines combined with either bone marrow or peripheral stem cell transplantation has enabled high dose chemotherapy to be administered [3]. Among the cytokines currently in clinical use are granulocyte colony-stimulating factor (G-CSF), interleukin-3 (IL-3) and granulocyte-macrophage colony stimulating factor (GM-CSF) [4, 5]. These measures besides being expensive offer short-term protection and do not eliminate myelosuppression particularly against multiple cycles of high-dose chemotherapy. To overcome chemotherapy-induced myelosuppression, transfer of drug resistant genes into hematopoietic cells may be a promising alternative or addition to stem cell transplantation combined with cytokine support [6–9]. Peripheral blood stem cells are easily harvested and reinfused without the need for any special measures to ensure homing to the marrow and engraftment. Potential advantages expected from gene transfer into hematopoietic progenitors include (1) elimination of postchemotherapy induced nadir; (2) may offer multilineage protection, and (3) may allow more frequent administration of high-dose chemotherapy at shorter intervals. In principle, the strategy involves introduction of drug resistance genes into hematopoietic progenitor cells ex vivo. These cells are then reinfused into the recipient prior to subsequent administration of high dose chemotherapy. This concept has been validated in several animal models using drug resistance genes, including mutant dihydrofolate reductase (DHFR) [10–14]; human multidrug resistance gene [15–18]; cytidine deaminase [19, 20] and DNA alkyl transferase and aldehyde dehydrogenase [21, 22]. Moreover, the concept that a higher dose of chemotherapeutic agent may lead to better tumor regression was recently demonstrated in a murine breast tumor model where high-dose methotrexate (MTX) administration after transplantation was curative in a significant number of tumor-bearing animals transplanted with bone marrow cells containing mutant DHFR cDNA [23]. Although several drug-resistant genes have been used to impart resistance, there are no reports available on the use of mutant thymidylate synthase (TS) genes to impart resistance to marrow progenitor cells in vitro or in vivo.

Thymidylate Synthase

TS is the rate-limiting step in the de novo synthesis of pyrimidines. The enzyme catalyzes the conversion of dUMP to dTMP by transfer of a methyl group from methylenetetrahydrofolate (cofactor) to the substrate (dUMP) (equation 1). The dTMP is then phosphorylated to dTTP and then incorporated into DNA:

$$dUMP + 5,10\text{-methylenetetrahydrofolate} \rightarrow dTMP + dihydrofolate \qquad (1)$$

Both MTX and 5-fluorouracil (5-FU) have been used in the clinic over the past three decades as anticancer agents. Several new anticancer agents that target TS have been synthesized and are going through the various stages of clinical development. These include raltitrexed (ZD1694;Tomudex®), AG337 (Thymitaq®), LY 231514, BW1843U89 [24]. A dose-limiting toxicity of these agents is bone marrow suppression, and therefore development of resistant variants of human TS will be useful in drug resistance gene transfer studies for protection against the myelosuppression induced by these agents.

Mutants of Human TS

The first human mutant TS cDNA to be identified and characterized is the Y33H mutant that confers modest resistance to fluoropyrimidines which

Table 1. Regions in human TS which are highly conserved through species and are important for ligand binding

Positions in human and *Lactophilus casei* TS	Sequence conservation among 29 TS species	Interactions with CH_2H_4 folate or dUMP
Phe225/Phe228	highly conserved with 1 exception: His	hydrophobic contact with PABA of folate
Leu221/Leu224	highly conserved with 1 exception: Val	hydrophobic contact with PABA of folate
Ile108/Ile81	highly conserved with 2 exceptions: Val and Tyr	hydrophobic contact with PABA of folate
Lys47/Lys20	invariant in vertebrates	Arg50 loop
Asp49/Asp22	strictly invariant	Arg50 loop
Arg50/Arg23	highly conserved with 1 exception: Gly	hydrogen bond with dUMP and C terminus
Thr51/Thr24	highly conserved with 1 exception: Gln	Arg50 loop
Gly52/His25	highly conserved with 5 exceptions: 2 His, Arg, Met and Pro	Arg50 loop
Phe59/Phe32	highly conserved with 2 exceptions: Met and Thr	β-sheet i, forming part of the substrate binding pocket
Gln214/Gln217	highly conserved with 1 exception: Ala	β-sheet iii, a kink region for three β-sheet formation

was developed in a cell line HCT116, made resistant to 5-FU by stepwise increments of the drug [25].

This mutant enzyme provides approximately 3-fold resistance to FdUrd and 3-fold resistance to Thymitaq [25, 26]. In order to develop more robust mutants of human TS, we have taken the following two approaches: (1) random ethylmethane sulfonate (EMS) mutagenesis and (2) a site-directed approach based on x-ray crystal information available for rat and bacterial TS enzymes in order to mutate specific sites.

Ethylmethane Sulfonate Mutagenesis

HT1080 human fibrosarcoma cells in logarithmic growth phase were exposed to 400 μg/ml of EMS for 18 h. The EMS-treated cells were then exposed to 40 μM Thymitaq, and surviving cells were cloned and established as cell lines. Forty-one colonies were obtained after EMS and Thymitaq exposure while only one colony was obtained from control cells treated with Thymitaq without prior EMS exposure. Analysis of these cell lines for mechanisms of resistance to Thymitaq indicated that a majority of them had increased levels of the TS protein without any evidence of mutations while some of them had point mutations in the TS gene with or without increased levels of TS protein. The following TS mutants were identified from the EMS studies: K47E, D49G, R50C, T51A, G52S, F59L and Q21R [27].

Site-Directed Mutagenesis

The random mutagenesis studies utilizing EMS-generated mutants around the substrate-binding region especially around the R50 amino acid. We reasoned that the region around the folate-binding sites of TS would also be interesting areas to mutate in order to generate drug-resistant TS proteins. Using X-ray crystal structure information available for bacterial and rat TS and applying them to human TS protein structure, we targeted three highly conserved regions around the folate binding site for human TS, i.e. the I108, the L221 and the F225 residues, and generated 14 different mutations [28]. Table 1 illustrates the regions in human TS which are highly conserved through species and are important for ligand binding.

Characterization of Mutant TS

In order to characterize these mutations with regard to the catalytic activity and drug resistance, we first tested the ability of these mutants to rescue TS lines after transfection and growth in media lacking thymidine. Only three of the mutants, i.e. K47E, D49G, G52S and the wtTS were able to complement growth of TS cells in selective medium lacking thymidine. The other 5 mutants were unable to rescue TS cells and were thus considered to be catalytically poor enzymes and not studied further. Several clones of each surviving transfectant were established as cell lines and examined for levels of TS protein and sensitivity to various anti-TS drugs such as Thymitaq, Tomudex, BW1843U89 and FdUrd. In order to rule out the possibility of overexpression of TS as a major mechanism of resistance to the anti-TS drugs, clones that expressed similar levels of TS protein as the wild-type (wt)TS transfectant were chosen for comparative growth inhibition studies. The results of the cytoxicity studies are presented in table 2 and it can be seen that mutants D49G and G52S conferred resistance to Thymitaq (40- and 12-fold, respectively) and to FdUrd (26- and 97-fold, respectively). No resistance was seen against Tomudex or BW1843U89. For the mutants arising out of the site directed

Table 2. Inhibition constants K_i (nM)

Mutants of TS	Raltitrexed	U89	AG337	FdUrd
Wild type	7.0	0.09	16	11
I108A	4,100	2,400	1,400	22
F225W	87	23	5.3	25
G52S	n.d.	n.d.	n.d.	n.d.
D49G	53	180	130	23

n.d. = Not determined.

Table 3. Kinetic parameters for wtTS and mutant human TS

Enzyme	K_m CH$_2$H$_4$F, μM	K_m dUMP, μM	K_{cat}, s^{-1}
wtTS	13	3.9	1.7
I108A	746	15	0.22
F225W	165	3.6	2.3
G52S	10	4.5	3.5
D49G	43	11	0.57

mutagenesis studies the I108A conferred resistance to Tomudex, Thymitaq and FdUrd (54-, 80- and 16-fold, respectively, over wtTS) while the F225W conferred 17-fold resistance to BW1843U89.

In order to determine the kinetic properties of the mutant enzymes these mutants were expressed in bacteria and purified. The kinetic properties of the mutant TS enzymes are shown in table 3.

Transfection of Mutant TS cDNA

The mutants G52S, I108A and F225W were chosen for gene transfer in order to impart drug resistance to transfected cells. As the first step toward bone marrow protection, these mutant cDNAs were cloned in the SFG retroviral vector [29]. Retrovirus producer lines were generated from the AM12 packaging cells and the titer for each producer line was determined using NIH3T3 cells as targets. Supernatants from these producer lines were used as the source of infectious particles for gene transfer into NIH 3T3 cells. A series of cytotoxicity assays were performed on the transduced cells using the various anti-TS drugs. The resistance profile of the different mutants to the drugs

Table 4. Resistance to Raltitrexed and Thymitaq in mouse CFU-GM assays[1]

Bone marrow transfected with	Colonies w/o drug	Colonies in 20 nM Raltitrexed	Colonies in 200 nM Thymitaq
No DNA	82	0	0
wtTS	108	0	0
I108A TS	144	14 (10%)	12 (8%)

[1] Transfection was carried out using the DOTAP transfection reagent from Boehringer Mannheim, Indianapolis, Ind., USA.

were consistent with the in vitro results obtained from both kinetic studies using the recombinant enzyme as well as the transfection studies using the TS cells. The next aim was to determine whether the mutant TS cDNAs imparted drug resistance to bone marrow cells in vitro by determining the growth of CFU-GM colonies in the presence and absence of the various drugs. Emergence of drug-resistant CFU-GM colonies indicated that the mutant TS cDNAs were indeed capable of imparting drug resistance to hematopoietic cells in vitro as shown in table 4.

Future Directions in Gene Transfer of Drug-Resistant Mutants of Human TS

The next set of experiments will be directed toward transplantation of bone marrow cells transduced with the mutant TS cDNA into irradiated recipients and challenging them with the appropriate drugs in order to determine the extent of myeloprotection. At the same time, we have also initiated work on generating transgenic animals expressing the mutant TS cDNAs in order to carry out more detailed toxicological analyses in these transgenic animals as well as bone marrow transplant recipients transplanted with marrow cells from these transgenic animals [O.A. O'Connor, unpubl. work in progress].

References

1   Hryniuk W, Busch H: The importance of dose intensity in chemotherapy of metastatic breast cancer. J Clin Oncol 1984;2:1281–1288.
2   Bierman PJ, Bagin RG, Jagannath S, Vose JM, Spitzer G, Kessinger A, Dicke KA, Armitage JO: High dose chemotherapy followed by autologous hematopoietic rescue in Hodgkin's disease: Long-term follow-up in 128 patients. Ann Oncol 1993;4:767.

3   Ellis GK, Hutchins L, Jimenez-Martin M, Pecora AL, Barnabas A, Meisenberg B, Nabholtz JM, Cortes-Funes H, Rifkin R, Chang AYC, Garrison L, George C, Giles FJ: Phase III double blind randomized study of PIXY321 versus G-CSF after CEP for breast or ovarian carcinoma. Blood 1996;88:449a.
4   Hofstra LS, Trope CG, Willemse PBH, Vindevoghel A, van den Bulche JM, Lahouseny M, Sklenar I, de Vries EGE: Randomized trial of rhIL-3 versus placebo in prevention of bone marrow depression during first-line chemotherapy for ovarian carcinoma. Proceedings ASCO 1997;16:115a.
5   Petzer AL, Zandstra PW, Piret JM, Eaves CJ: Differential cytokine effects on primitive (CD34+CD38−) human hematopoietic cells: Novel responses to Flt-3 ligand and thrombopoietin. J Exp Med 1996;183:2551–2558.
6   Banerjee D, Zhao SC, Li M-X, Schweitzer BI, Mineishi S, Bertino JR: Gene therapy utilizing drug resistant genes. A review. Stem Cells 1994;12:378–384.
7   Miller AD: Progress toward human gene therapy. Blood 1990;76:271–278.
8   Koc ON, Allay JA, Lee K, Davis BM, Reese JS Gerson SL: Transfer of drug resistance genes into hematopoietic progenitors to improve chemotherapy tolerance. Semin Oncol 1996;23:46–65.
9   Bertino JR: Ode to methotrexate. J Clin Oncol 1993;11:5–14.
10 Williams DA, Hsieh K, DeSilva A, Mulligan RC: Protection of bone marrow transplant recipients from lethal doses of methotrexate by the generation of methotrexate resistant bone marrow. J Exp Med 1987;166:210–218.
11 Corey CA, DeSilva AD, Holland CA, Williams DA: Serial transplantation of methotrexate resistant bone marrow; protection of murine recipients from drug toxicity by progeny of transduced stem cells. Blood 1990;76:337–343.
12 May C, Gunther R, McIvor RS: Protection of mice from lethal doses of methotrexate by transplantation with transgenic bone marrow expressing drug resistant dihydrofolate reductase activity. Blood 1995;86:2439–2448.
13 Zhao SC, Li M-X, Banerjee D, Schweitzer BI, Mineishi S, Gilboa E, Bertino JR: Long term protection of recipient mice from lethal doses of methotrexate by marrow infected with a double copy vector retrovirus containing a mutant dihydrofolate reductase. Cancer Gene Ther 1994;1:27–33.
14 Li M-X, Banerjee D, Zhao SC, Schweitzer BI, Mineishi S, Gilboa E, Bertino JR: Development of a retroviral construct containing a human mutated dihydrofolate reductase cDNA for hematopoietic stem cell transduction. Blood 1994;83:3403–3408.
15 Sorrentino BP, Brandt SJ, Bodine D, Gottesman M, Pastan I, Cline A, Nienhuis AW: Selection of drug resistant bone marrow cells in vivo after retroviral transfer of human MDR1. Science 1992;257:99–103.
16 Podda S, Ward M, Himelstein A, Richardson C, de la Flor-Weiss E, Smith L, Gottesman M, Pastan I: Transfer and expression of the human multiple drug resistance gene into live mice. Proc Natl Acad Sci USA 1992;89:9676–9680.
17 Gottesman MM, Germann UA, Aksentijevich I, Sugimoto Y, Cardarelli CO, Pastan I: Gene transfer of drug resistant genes. Implications for cancer therapy. Ann NY Acad Sci 1994;716:126–138.
18 Ward M, Richardson C, Pioli P, Smith L, Podda S, Goff S, Hesdorffer C, Bank A: Transfer and expression of the human multiple drug resistance gene in human CD34+ cells. Blood 1994;84:1408–1414.
19 Momparler RL, Eliopoulos N, Bovenzi V, Letourneau S, Greenbaum M, Cournoyer D: Resistance to cytosine arabinoside by retrovirally mediated gene transfer of human cytidine deaminase into murine fibroblast and hematopoietic cells. Cancer Gene Ther 1996;3:331–338.
20 Neff T, Blau CA: Forced expression of cytidine deaminase confers resistance to cytosine arabinoside and gemcitabine. Exp Hematol 1996;24:1340–1346.
21 Moritz T, Mackay W, Glassner BJ, Williams DA, Samson L: Retrovirus mediated expression of a DNA repair protein in bone marrow protects hematopoietic cells from nitrosourea induced toxicity in vitro and in vivo. Cancer Res 1995;55:2608–2611.
22 Magni M, Shammah S, Schiro R, Mellado W, Dalla-Favera R, Gianni AM: Induction of cyclophosphamide resistance by aldehyde dehydrogenase gene transfer. Blood 1996;87:1097–1103.
23 Zhao SC, Banerjee D, Mineishi S, Bertino JR: Post transplant methotrexate administration leads to improved curability of mice bearing a mammary tumor transplanted with marrow transduced with a mutant human dihydrofolate reductase cDNA. Hum Gene Ther 1997;8:903–909.

24 Bertino JR, Li WW, Lin J, Trippett T, Goker E, Schweitzer BI, Banerjee D: Enzymes of the thymidylate cycle as targets for chemotherapeutic agents. Mechanisms of resistance. Mt Sinai J Med 1992;59:391–395.

25 Hughey CT, Barbour KW, Berger FG, Berger SH: Functional effects of a naturally occurring amino acid substitution in human thymidylate synthase. Mol Pharmacol 1993;44:316–323.

26 Frantz C Spencer HT: Retroviral gene transfer of a thymidylate synthase-dihydrofolate reductase chimera confers fluoropyrimidine and antifolate resistance. 39th Annual Meeting, American Society of Hematology, San Diego. Blood 1997;90:4599a.

27 Tong Y, Chen X-L, Ercikan-Abali EA, Capiaux GM, Zhao SC, Banerjee D, Bertino JR: Isolation and characterization of Thymitaq and 5-fluoro-2-deoxyuridylate (FdUMP)- resistant mutants of human thymidylate reductase from ethylmethane sulfonate-exposed human sarcoma HT1080 cells. J Biol Chem 1998;273:11611–11618.

28 Tong Y, Chen X-L, Ercikan-Abali EA, Capiaux GM, Zhao SC, Banerjee D, Maley F, Bertino JR: Probing the binding site of human thymidylate synthase by site directed mutagenesis. Evidence that Tomudex, Thymitaq and BW1843U89 bind differently to this enzyme. J Biol Chem 1998;273: 31209–31214.

29 Riviere I, Sadelain M: Methods for the construction of retroviral vectors and the generation of high-titer producers; in Robbins P (ed): Methods in Molecular Medicine. Gene Therapy Protocols. Totowa, Humana Press, 1997, pp 59–78.

D. Banerjee, Box 78, MSKCC, 1275 York Avenue, New York, NY 10021 (USA)
Tel. +1 212 639 8307, Fax +1 212 639 2767, E-Mail banerjed@mskcc.org

Dihydropyrimidine Dehydrogenase and Resistance to 5-Fluorouracil

Robert B. Diasio

Department of Pharmacology and Toxicology, Division of Clinical Pharmacology, University of Alabama at Birmingham, Birmingham, Ala., USA

Although originally synthesized 40 years ago, the cancer chemotherapy drug 5-flourouracil (5-FU) continues to be widely used in the treatment of several of the common human malignancies including carcinoma of the colon, breast and skin [1]. This drug belongs to the antimetabolite class of chemotherapy agents. It is an analog of the pyrimidine uracil and as such is taken up into the cell and metabolized via both the anabolic and catabolic pathways similar to uracil [2]. Today there are a number of fluoropyrimidines that are used as chemotherapy drugs. These include 5-FU and its deoxyribonucleoside, 5-fluoro-2-deoxyuridine (FdUrd) as well as several 5-FU prodrugs (slow-release 5-FU).

The fluoropyrimidine drugs enter the pyrimidine metabolic pathways where they are anabolized to nucleotides that interfere or block normal nucleic acid formation. Anabolism has been the major site for biochemical and molecular investigations over the past 4 decades, as it is recognized that these drugs must first be anabolized before antitumor activity is obtained. Figure 1 illustrates the major metabolic pathways and demonstrates the three major hypothesized sites of action: (1) formation of fluorodeoxyuridine monophosphate (FdUMP) which can complex with thymidylate synthase in the presence of 5,10-methylene tetrahydrofolate to inhibit the formation of thymidylate needed for DNA synthesis; (2) formation of fluorouridine triphosphate (FUTP) which can be incorporated into newly synthesized RNA resulting in RNA dysfunction, and (3) formation of fluorodeoxyuridine triphosphate (FdUTP) which can be incorporated into newly synthesized DNA resulting in DNA fragmentation.

As shown in figure 1, the fluoropyrimidine drugs can also enter the pyrimidine catabolic pathway as 5-FU and be converted to metabolites corresponding

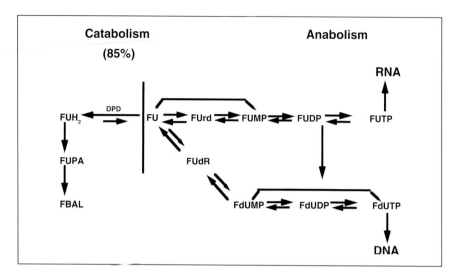

Fig. 1. Metabolic overview illustrating the critical position of DPD in the metabolism of 5-FU as well as the natural pyrimidines uracil and thymine. More than 85% of administered 5-FU is catabolized via DPD.

to the naturally occurring catabolites of uracil. In contrast to anabolism, there has been much less attention focused on catabolism over the past 40 years. Dihydropyrimidine dehydrogenase (dihydrouracil dehydrogenase, dihydrothymine dehydrogenase, uracil reductase; EC 1.3.1.2, DPD) is the initial rate-limiting enzymatic step in the catabolism of not only the naturally occurring pyrimidines uracil and thymine, but also of 5-FU [1, 3]. DPD thus occupies an important position in the overall metabolism of 5-FU being responsible for converting over 85% of clinically administered 5-FU to 5-FUH$_2$ [4], an inactive metabolite, in an enzymatic step that is essentially irreversible (fig. 1). While anabolism is clearly essential for the antitumor activity of 5-FU, catabolism, by controlling the availability of 5-FU for anabolism, indirectly is also a critical determinant of the antitumor activity of 5-FU.

Importance of Dihydropyrimidine Dehydrogenase for 5-Fluorouracil Pharmacology

The importance of DPD to the clinical pharmacology of 5-FU has been further emphasized by several recent studies that demonstrate how DPD can influence the pharmacokinetics, bioavailability, toxicity, and antitumor effec-

Table 1. Importance of DPD in 5-FU pharmacology

Circadian variation of DPD responsible for circadian variation of 5-FU during 5-FU protracted infusion

Variation of DPD (in population) responsible for variability in 5-FU pharmacokinetics ($t_{\frac{1}{2}}$ and clearance)

Variation nof intestinal DPD (in population) responsible for variability in 5-FU bioavailability (oral administration)

Pharmacogenetic syndrome – DPD deficiency – Small percent (<3%) of population at risk

Increased levels of DPD in tumor tissue may be basis for 5-FU resistance

Inhibitors of DPD alter 5-FU pharmacology

---

tiveness of 5-FU as well as its use with other drugs [5]. These are summarized in table 1.

DPD is now known to follow a circadian pattern in both animals and humans [6–8]. Studies in patients receiving 5-FU infusion by automated pumps have demonstrated that the circadian variation of tissue DPD level is accompanied by an inverse circadian patterns in plasma 5-FU concentrations. This has potential importance in the design of time-modified 5-FU infusions. Such regimens have been suggested to have potential benefit in the treatment of certain human cancers [9].

DPD enzyme activity in normal tissues (peripheral blood mononuclear cells and liver) has also been shown to vary from individual to individual in a Gaussian pattern with as much as a 6-fold variation from the lowest to the highest values [10, 11]. This wide variation in DPD activity is likely responsible for the wide variation in the $t_{\frac{1}{2}\beta}$ and clearance of 5-FU observed in pharmacokinetic studies [4].

In addition to the variation of DPD activity in the normal population, there is a small percentage (<3%) of the population that has DPD activity significantly below (>2 SD below mean) the Gaussian distribution that characterizes most of the population [12, 13]. These individuals are at significant risk if they develop cancer and are given 5-FU [14]. Thus, this is a true pharmacogenetic syndrome, with symptoms not being recognized until exposure to the drug [15].

The variation in DPD activity has also recently been shown to be responsible for the apparently variable bioavailability of orally administered 5-FU. The basis for the erratic bioavailability of orally administered 5-FU has not previously been clear, particularly since 5-FU is a small molecule with a pK_a that should result in excellent absorption and bioavailability. Experimental studies in animals using DPD inhibitors have demonstrated that following inhibition of DPD the pharmacokinetic pattern resulting from oral administration of 5-FU is essentially the same as that produced by intravenous administration resulting in essentially 100% bioavailability [16].

Tumors may also express a variable level of DPD activity [17]. This may explain the observed varied tumor response to 5-FU, and thus be a basis for resistance to 5-FU (see below). Thus tumors with high DPD levels are predicted to be relatively resistant to 5-FU, while tumors with low levels of DPD would be expected to be relatively sensitive to 5-FU.

It should be noted that over the years there has been a concerted effort to intentionally develop inhibitors or DPD as a means of increasing the effect of 5-FU [18]. The rationale for developing these inhibitors was based on the realization that it was essential for 5-FU be anabolized to 5-FU nucleotides in order for antitumor activity to occur, while at the same time appreciating that most of the administered 5-FU was catabolized [2]. Thus it was felt by experimental chemotherapists that it was desirable to inhibit catabolism to increase the anabolism and hence the antitumor effect. Unfortunately, with many of the initial inhibitors, there has been marked host toxicity. Recently, there have been attempts to utilize new inhibitors of DPD not only to increase the anabolism of 5-FU and hence the antitumor activity, but also to achieve desirable pharmacologic effects, e.g. improve the bioavailability of 5-FU (as described above). One such example of a therapeutically useful inhibitor of DPD is ethynyluracil (GW 776C85) which has been shown to improve 5-FU efficacy and selected pharmacologic effects (e.g. bioavailability) in both preclinical and early clinical studies [19, 20]. Several additional DPD inhibitors are now being evaluated with 5-FU in clinical trials in an attempt to achieve similar effects.

Resistance to 5-Fluorouracil

As with other cancer chemotherapy drugs, there are a variety of resistance mechanisms (both innate and acquired) by which tumor cells may become resistant to 5-FU. Knowledge of the metabolism of 5-FU, its mechanisms of action, and other factors (including cofactors) that can influence the concentration and duration of exposure of active metabolites at the sites of action permits one to propose possible means by which resistance can develop to 5-FU. Resistance may theoretically develop at any of the various steps in the passage of 5-FU through the cell from initial entry across the cell membrane through the various metabolic steps to eventually interaction with the intracellular target(s) responsible for drug action [2, 21, 22].

Although it remains a theoretical possibility that an alternation of the 'transport' of 5-FU into cells may be a mechanism of de novo or acquired resistance, there is no convincing evidence that such a mechanism exists. This is in contrast to the pleotropic or multidrug resistance mechanism that characterizes drug resistance with many other antineoplastic drugs. This latter type of resistance utilizes a pumping mechanism present in the cell membrane to pump or efflux drug (even against a gradient) from the cell [23]. In contrast to most of these drugs which include a number of diverse chemical structures representing compounds which are recognized by the cells as 'foreign', 5-FU has a structure that is similar to natural chemical structures utilized in nucleic acid synthesis and thus is not recognized as foreign by this pump.

The search for biochemical and molecular mechanisms of 5-FU resistance has essentially paralleled our understanding of the metabolism and mechanism of action of 5-FU. It was appreciated early that primary or secondary resistance was often a reflection of decreased conversion of 5-FU to nucleotides [24], leading to suggestions that one might measure the conversion of 5-FU to nucleotides in human tissues as a means to predict resistance. Subsequently, there was interest in the levels of anabolic enzymes responsible for conversion of 5-FU to nucleotides, particularly uridine phosphorylase and uridine kinase, as well as orotate phosphoribosyltransferase used to convert 5-FU to fluorouridine monophosphate (FUMP) [25]. Similarly, there was interest in deoxyuridine phosphorylase and deoxyuridine kinase not only with 5-FU but also with FdUrd resistance. It soon became clear that other factors, such as the availability of cofactors (eg. PRPP or ATP) could also account for resistance [2]. Theoretically, resistance can also occur at any of the enzymatic steps (fig. 1) involved in the formation of FUMP, FUTP, FdUMP, or FdUTP as well as the incorporation of FdUTP and FUTP into DNA or RNA, respectively. It should be noted that repair mechanisms at the level of DNA or RNA or enzymes such as dUTPase which can convert FdUTP to FdUMP decreasing the likelihood of 5-FU incorporation into DNA have also been proposed as potential mechanisms of resistance [2].

Because of its critical importance in the cytotoxicity of 5-FU, much attention has been focussed over the past few decades on thymidylate synthase (TS) as a site of resistance [21, 22]. Since inhibition of TS requires sufficient quantities of FdUMP and 5,10-methylene tetrahydrofolate to form a ternary complex that is critical for the inhibition to occur, lack of either or both of these has been the focus of attention for those studying resistance to 5-FU [21, 22]. More recently, it has been appreciated that the increased TS levels, possibly by gene amplification, can also be a source of resistance [26].

While all of the above examples of resistance focussed on steps in the anabolic pathway, it is clear from recent studies that catabolism can also contribute to 5-FU resistance. Thus, theoretically, if catabolism of 5-FU were to increase within a tumor, one would expect increased degradation with less 5-FU available for anabolism to the critical levels of 5-FU nucleotides needed at the target sites to produce antitumor activity. Of particular interest is evidence that demonstrates that this mechanism of resistance is in fact present clinically.

Role of Dihydropyrimidine Dehydrogenase in Resistance to 5-Fluorouracil

As noted above, within the catabolic pathway, DPD is the rate-limiting step. DPD thus is not only the key enzyme in pyrimidine catabolism, but also has a critical position in the overall metabolism of 5-FU, regulating the total amount of 5-FU available for anabolism. The variability of DPD within host tissues [11] such as liver and gastrointestinal mucosa can account indirectly for slight variations in the overall exposure to 5-FU. This may theoretically contribute to resistance. Thus individuals with increased liver DPD activity may be expected to catabolize 5-FU relatively rapidly and thereby be relatively resistant not only to 5-FU toxicity but also 5-FU antitumor activity.

More recently, some tumors have been demonstrated to have relatively increased DPD levels that may be more directly the cause of 5-FU resistance [27]. It is interesting that some tumors (e.g. hepatoma) that are relatively resistant to 5-FU or its derivatives have relatively high DPD levels [17].

While DPD enzyme activity within tumors can be measured to assess de novo or acquired resistance in some clinical specimens, it is not practical for most samples. We have recently utilized PCR methodology to examine levels of DPD mRNA in tumor and observed that resistant tumors often have increased levels of DPD mRNA [28]. The use of PCR makes this method useful for small tissue samples such as that obtained from a needle biopsy. We have used this quantitative PCR methodology to analyze 33 colorectal cancer specimens collected from patients at the time of surgery who subsequently were treated with 5-FU [29]. DPD mRNA levels were found to vary over 80-fold in these tissue samples. DPD expression in the subgroup of 11 patients who responded to 5-FU had values (DPD/β-actin ratios) of 2.5 or less. DPD expression in the subgroup of 22 patients who were resistant to 5-FU had values (DPD/β-actin ratios) of greater than 2.5, with the highest value being 16. The median survival of patients with values >2.5 was 5 months compared to 10 months for those with values <2.5 (p=0.0015). Of particular interest is the assessment of other important 5-FU-metabolizing enzymes together with DPD in predicting resistance. In the above study, the use of quantitation of DPD expression together with thymidylate synthase expression leads to prediction of all 11 of the 5-FU responders and 21/22 of the 5-FU-resistant patients.

Overcoming 5-Fluorouracil Resistance by Pharmacologic Modulation of Dehydropyrimidine Dehydrogenase

It becomes attractive to consider inhibiting DPD in order to eradicate the variability in 5-FU pharmacology. Inhibiting DPD in 5-FU-susceptible host tissue, such as gastrointestinal mucosa and bone marrow, should make dosing from patient to patient less variable with this cancer chemotherapeutic agent in which dosing decisions are typically based on observed toxicity. Inhibition of DPD activity directly in the tumor is also attractive, particularly since it is likely that many tumors become 5-FU resistant through an increase in intratumoral DPD activity resulting in increased degradation and thus less anabolism of 5-FU.

Future studies are needed to determine whether effective DPD inhibitors such as ethynyluracil (GW776C85) can lead to improved 5-FU therapy, particularly by overcoming the increased DPD levels in tumors that are likely the basis for 5-FU resistance.

Conclusions

Resistance to 5-FU may theoretically develop at any of the various steps in the passage of 5-FU through the cell from initial entry across the cell membrane through the various metabolic steps to eventual interaction with the intracellular target(s) responsible for drug action. In the past several years, DPD also has been recognized as a potential site of resistance. This is understandable since DPD is a critical step in pyrimidine metabolism. Most convincing is the demonstration of increased levels of DPD in tumors of patients who are clinically resistant to 5-FU. The recent development of DPD inhibitors offers the possibility of overcoming the resistance due to increased DPD levels in the tumors.

References

1 Diasio RB, Harris BE: Clinical pharmacology of 5-fluorouracil. Clin Pharmacokinet 1989;16: 215–237.
2 Daher GC, Harris BE, Diasio RB: Metabolism of pyrimidine analogues and their nucleosides; in Powis G (ed): Anticancer Drugs: Antimetabolite Metabolism and Natural Anticancer Agents. Oxford, Pergamon Press, 1994, chapt 2.
3 Lu Z, Zhang R, Diasio RB: Purification and characterization of dihydropyrimidine dehydrogenase from human liver. J Biol Chem 1992;267:17102–17109.
4 Heggie GD, Sommadossi JP, Cross DS, Huster WJ, Diasio RB: Clinical pharmacokinetics of 5-fluorouracil and its metabolites in plasma, urine, and bile. Cancer Res 1987;47:2203–2206.
5 Diasio RB, Lu Z: Dihydropyrimidine dehydrogenase activity and 5-fluorouracil chemotherapy. J Clin Oncol 1994;12:2239–2242.
6 Harris BE, Song R, He Y, Soong S-J, Diasio RB: Circadian rhythm of rat liver dihydropyrimidine dehydrogenase: Possible relevance to fluoropyrimidine chemotherapy. Biochem Pharmacol 1988;37: 4759–4763.
7 Harris BE, Song R, Soong S-J, Diasio RB: Circadian variation of 5-fluorouracil catabolism in isolated perfused rat liver. Cancer Res 1989;49:6610–6614.
8 Harris BE, Song R, Soong S-J, Diasio RB: Relationship of dihydropyrimidine dehydrogenase activity and plasma 5-fluorouracil levels: Evidence for circadian variation of 5-fluorouracil levels in cancer patients receiving protracted continuous infusion. Cancer Res 1990;50:197–201.
9 Zhang R, Diasio RB: Pharmacologic basis for circadian pharmacodynamics; in Hrushesky WJM (ed): The Scientific Basis for Optimized Cancer Therapy. Boca Raton, CRC Press, 1994, chapt 4.
10 Lu Z, Zhang R, Diasio RB: Dihydropyrimidine dehydrogenase activity in human peripheral blood mononuclear cells and liver: Population characteristics, newly identified patients, and clinical implication in 5-fluorouracil chemotherapy. Cancer Res 1993;53:5433–5438.
11 Lu Z, Zhang R, Diasio RB: Dihydropyrimidine dehydrogenase activity in human liver: Population characteristics and clinical implication in 5-FU chemotherapy. Clin Pharmacol Ther 1995;58:512–522.
12 Diasio RB, Beavers TL, Carpenter JT: Familial deficiency of dihydropyrimidine dehydrogenase: Biochemical basis for familial pyrimidinemia and severe 5-fluorouracil-induced toxicity. J Clin Invest 1988;81:47–51.
13 Harris BE, Carpenter JT, Diasio RB: Severe 5-fluorouracil toxicity secondary to dihydropyrimidine dehydrogenase deficiency: A potentially more common pharmacogenetic syndrome. Cancer 1991; 68:499–501.
14 Takimoto CH, Lu Z, Zhang R, Liang MD, Larson LV, Cantilena LR Jr, Grem JL, Allegra CJ, Diasio RB, Chu E: Severe neurotoxicity following 5-fluorouracil-based chemotherapy in a patient with dihydropyrimidine dehydrogenase deficiency. Clin Cancer Res 1996;2:477–481.
15 Lu Z, Diasio RB: Polymorphic drug metabolizing enzymes; in Schilsky RL, Milano GA, Ratain MJ (eds): Principles of Antineoplastic Drug Development and Pharmacology. New York, Marcel Dekker Inc, 1996.
16 Baccanari DP, Davis ST, Knick VC, Spector T: 5-Ethynyluracil: Effects on the pharmacokinetics and antitumor activity of 5-fluorouracil. Proc Natl Acad Sci USA 1993;90:11064–11068.
17 Jiang W, Lu Z, He Y, Diasio RB: Dihydropyrimidine dehydrogenase activity in hepatocellular carcinoma; implication for 5-fluorouracil-based chemotherapy. Clin Cancer Res 1997;3:395–399.
18 Naguib FNM, el Kouni MH, Cha S: Enzymes of uracil catabolism in normal and neoplastic human tissues. Cancer Res 1985;45:5405–5412.
19 Cao S. Rustum YM, Spector T: 5-Ethynyluracil (776C85); modulation of 5-fluorouracil efficacy and therapeutic index in rats bearing advanced colorectal cancer. Cancer Res 1994;54:1507–1510.
20 Baker SD, Khor SP, Adjei AA, Doucette M, Spector T, Donehower RC, Grochow LB, Sartorius SE, Noe DA, Hohneker JA, Rowinsky EK: Pharmacokinetics, oral bioavailability and safety study of fluorouracil in patients treated with 776C85, an inactivator of dihydropyrimidine dehydrogenase. J Clin Oncol 1996;14:3085–3096.

21 Spears CP: Clinical resistance to antimetabolites. Hematol Oncol Clin North Am 1995;9:397–413.
22 Chu E, Allegra C: Mechanisms of clinical resistance to 5-fluorouracil chemotherapy. Cancer Treat Res 1996;87:175–195.
23 Endicott JA, Ling V: The biochemistry of P-glycoprotein-mediated multidrug resistance. Annu Rev Biochem 1989;58:137–171.
24 Kessel D, Hall TC, Wodinsky I: Nucleotide formation as a determinant of 5-fluorouracil response in murine leukemias. Science 1966;154:911–913.
25 Goldberg AR, Machledt JH, Pardee AB: On the action of fluorouracil on leukemia cells. Cancer Res 1966;26:1611–1615.
26 Clark JL, Berger SH, Mittelman A, Berger FG: Thymidylate synthase gene amplification in a colon tumor resistant to fluoropyrimidine chemotherapy. Cancer Treat Rep 1989;71:261–265.
27 Milano G, Etienne MC, Fischel JL: Intratumoral dihydropyrimidine dehydrogenase and its inhibition: A new approach in the pharmacokinetics of 5-fluorouracil. Ann Gastroentérol Hépatol 1995;31:103–105.
28 Diasio RB, Danenberg K, Johnson M, Danenberg P: Quantitation of dihydropyrimidine dehydrogenase expression in tumor specimens of patients treated with 5-fluorouracil using a quantitative PCR assay. Proc Am Assoc Amer Cancer Res 1998;39:188.
29 Danenberg K, Salonga D, Park JM, Leichman CG, Leichman L, Johnson M, Diasio RB, Danenberg P: Dihydropyrimidine dehydrogenase and thymidylate synthase gene expressions identify a high percentage of colorectal tumors responding to 5-fluorouracil. Proc Am Soc Clin Oncol 1998;17:258.

Robert B. Diasio, MD, Department of Pharmacology and Toxicology, Box 600, Volker Hall, University of Alabama at Birmingham, Birmingham, AL 35294 (USA)
Tel. +1 205 934 4578, Fax +1 205 934 8240, E-Mail Robert.Diasio@ccc.uab.edu

Chemoprotection against Cytosine Nucleoside Analogs Using the Human Cytidine Deaminase Gene

Nicoletta Eliopoulos, Christian Beauséjour, Richard L. Momparler

Département de pharmacologie, Université de Montréal, Centre de recherche pédiatrique, Hôpital Ste-Justine, Montréal, Québec, Canada

Present-day chemotherapy for advanced metastatic cancer is not very effective [1]. The long-term survival of patients with metastatic lung, breast, prostate, and colon cancer treated with chemotherapy is less than 10%. New approaches to improve the therapeutic effectiveness should be explored. In general, escalation of the dose of antineoplastic agents, and/or frequency of treatment used in patients is required to obtain the maximum response. However, dose intensification for most anticancer drugs is limited by their toxicity to the bone marrow. A possible way to overcome this problem would be to render normal hematopoietic cells resistant to chemotherapeutic drugs and hence reduce their toxicity. Several investigators have demonstrated that gene transfer of drug resistance genes into marrow cells decreased their sensitivity to some classes of anticancer agents. For example, chemoprotection against different types of antineoplastic drugs has been well documented for the multidrug resistance (MDR) and mutant dihydrofolate reductase (DHFR) genes [2–5].

The potential effectiveness in cancer therapy, of cytosine nucleoside analogs, such as cytosine arabinoside (ARA-C), 2′,2′-difluorodeoxycytidine (dFdC), and 5-aza-2′-deoxycytidine (5-AZA-CdR), is reduced by their dose-limiting myelosuppression [6–8]. The cytidine deaminase (CD) gene which inactivates cytosine nucleoside analogs by deamination [9, 10] may be an interesting gene to confer resistance to these drugs. Gene therapy with the CD gene may have the potential to circumvent the hematopoietic toxicity produced by chemotherapy with cytosine nucleoside analogs and consequently, increase their clinical efficacy. This chapter contains data to support this hypothesis.

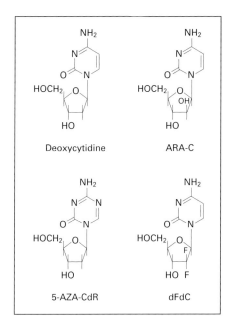

Fig. 1. Chemical structure of deoxycytidine and related analogs.

Potential Role of Cytosine Nucleoside Analogs in Tumor Therapy

Cytosine Arabinoside

Cytosine arabinoside (ARA-C) is one of the most studied nucleoside analogs. It differs structurally from the natural occurring, deoxycytidine, by the existence of a 2'-hydroxyl group in the trans position of the sugar moiety (fig. 1). For review, see Grant [11]. A characteristic of all cytosine nucleoside analogs, ARA-C is a prodrug that in order to be rendered a cytotoxic agent must first be phosphorylated by deoxycytidine kinase [12]. The incorporation of ARA-C into the 3'-terminal position of the DNA strand can lead to chain termination [11, 13]. ARA-C exerts its toxic effect on cells present in the S phase of the cell cycle [14].

ARA-C is one of the most active drugs used in the treatment of patients with acute leukemia [15, 16]. It also possesses activity against lymphomas [17, 18]. Although in vitro studies demonstrated antineoplastic activity of ARA-C against human tumor cell lines [19], this drug at various dose schedules tested has not been shown to be very effective in treating solid tumors [20]. However, in a recent report, a prolonged remission was achieved in a woman with advanced metastatic breast cancer subsequent to the administration of high-dose ARA-C [6].

2',2'-Difluorodeoxycytidine

dFdC is a cytosine nucleoside analog which contains two fluorine molecules at the 2'-position of the sugar ring (fig. 1). For review, see Parkinson et al. [21]; Plunkett et al. [22]. dFdC diphosphate exerts an inhibitory effect on ribonucleotide reductase. As such, the size of the dCTP pool is reduced by dFdC, thus potentiating its antineoplastic effect by decreasing the competition between the natural nucleotide and dFdC triphosphate for DNA polymerase [22]. DNA synthesis inhibition arises due to the incorporation of dFdC into DNA. An interesting phenomenon that occurs with dFdC, 'masked chain termination', is that after incorporation of this analog into DNA, DNA polymerase adds one additional deoxynucleotide. Consequently, since dFdC occupies a nonterminal spot in the DNA, proof reading exonucleases cannot excise it, resulting in a block of both DNA repair and DNA replication [22–24].

In clinical trials, dFdC therapy has exhibited promising activity against the following cancers: breast [25], pancreatic [26, 27], bladder [28], ovarian [29, 30], lung [7, 31, 32] and head and neck [33].

5-Aza–2'-Deoxycytidine

The cytosine nucleoside analog 5-AZA-CdR contains a nitrogen at the 5-position of the pyrimidine ring in place of a carbon (fig. 1). For review, see Momparler [34]. The active triphosphate form of this analog, 5-AZA-dCTP, is a good substrate for DNA polymerase and is readily incorporated into replicating DNA [34]. The presence of 5-AZA-CdR at specific positions in DNA produces a block in the methylation of cytosine residues in DNA [35]. The consequences of DNA demethylation are the transcriptional activation of genes which were silenced by methylation of CpG islands [36, 37]. In some cell types, this demethylation results in the induction of cellular differentiation [35, 38]. Recently, Bender et al. [39] demonstrated that 5-AZA-CdR produced a prolonged growth-inhibitory effect, even after drug removal, on several human tumor cell lines, possibly due to the reactivation, by demethylation, of growth regulatory genes.

In clinical studies, 5-AZA-CdR displayed favorable activity against childhood and adult leukemia [40–42], and myelodysplastic syndrome [48]. In a recent pilot trial by Momparler et al. [8], 5-AZA-CdR showed interesting activity against stage IV non-small cell lung cancer. Notably, one patient, the one that received 5 cycles of 5-AZA-CdR, more than any other patient, survived over 6 years after therapy. Significant interest in 5-AZA-CdR is being generated by investigators reporting on its activation of tumor suppressor genes that have been silenced by DNA methylation. These genes include VHL in renal carcinoma [44], retinoic acid receptor beta in colon cancer [45, 46] and p16 in lung tumors [47, 48], gliomas [49] and bladder cancer [50]. In

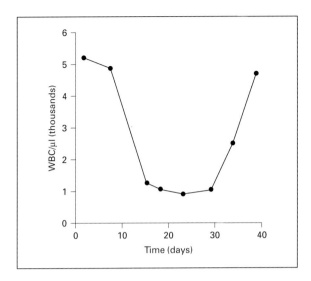

Fig. 2. Example of the hematopoietic toxicity in a patient with metastatic lung cancer following treatment with an 8-hour infusion of 5-AZA-CdR at a total dose of 660 mg/m^2 [8].

addition, 5-AZA-CdR can activate the expression of a gene responsible for the suppression of tumor cell invasion, E-cadherin, in primary tumors of breast and prostate [51]. Pilot clinical trials on 5-AZA-CdR are under investigation in patients with melanoma, breast and lung cancer [8].

Hematopoietic Toxicity Produced by Cytosine Nucleoside Analogs

Hematopoietic toxicity is a major and common complication of cancer chemotherapy [52]. The myelosuppression produced by most anticancer drugs, including cytosine nucleoside analogs [6–8], is dose-limiting and prevents dose escalation to improve clinical efficacy. Neutropenia is the most serious side effect observed following intensive treatment with cytosine nucleoside analogs and may result in a high risk of fatal infections.

S-phase-specific drugs, such as cytosine nucleoside analogs, produce a reduction of the neutrophil count, with a return to normal values, which varies in duration depending on the analog used. ARA-C, given as a continuous infusion at the conventional dose for 5 days, caused a leukopenia from about day 14 to 21 [20]. For dFdC, administered as short intravenous infusions (30–60 min), a granulocytopenia occurred with a nadir at day 15 [7]. Treatment with an 8-hour infusion of 5-AZA-CdR resulted in a leukopenia from day 15 to 30 (fig. 2) [8].

Chemoprotection against Hematopoietic Toxicity Using Drug Resistance Genes

Rationale

For many anticancer agents, including cytosine nucleoside analogs, any reduction in dose intensity can diminish the likelihood of cure due to insufficient cytotoxic concentrations of drug reaching the target tumor [53]. An innovative solution for guarding against the toxic side effects of chemotherapy would be to use gene therapy to render normal cells resistant to antineoplastic drugs hence permitting dose intensification in the absence of myelosuppression [54]. One advantage of chemoprotection is that only a single transplantation of gene-modified hematopoietic cells would be sufficient to confer protection against drug-induced hematopoietic toxicity for many subsequent cycles of drug treatment.

This goal of enhancing the tolerance of the hematopoietic system to the toxic effects of specific anticancer agents may be accomplished by using an efficient vector for transfer and expression of genes for drug resistance in normal hematopoietic cells [54–57]. Extensive preclinical studies have established that the expression of the drug resistance genes MDR and DHFR, in hematopoietic cells confers chemoprotection against the myelosuppression produced by MDR drugs and folate analogs, respectively [2–5]. More details on different drug resistance genes are discussed in other chapters of this volume.

The future clinical potential of utilizing the chemoprotection approach is illustrated in a study on mice with mammary adenocarcinoma which showed that total tumor regression was produced, after high-dose methotrexate therapy, in 44% of the mice reconstituted with marrow cells containing the DHFR transgene [58]. Various clinical trials have commenced on MDR1 gene transfer to hematopoietic cells in patients with advanced breast and ovarian cancer [59–62].

Chemoprotection against Cytosine Nucleoside Analogs

Current chemotherapy for advanced metastatic cancer is often not very effective [1]. The cytosine nucleoside analogs ARA-C, dFdC and 5-AZA-CdR show promising antitumor activity in patients with advanced lung and breast cancer [6–8]. Since more intensive chemotherapy with cytosine nucleoside analogs can lead to a better clinical response, their curative potential should be improved if their dose-limiting side effect of myelosuppression could be prevented. To achieve this, one approach would be to render normal bone marrow cells resistant to the cytotoxicity of cytosine nucleoside analogs by introducing into these cells the drug resistance gene, CD. The main purpose

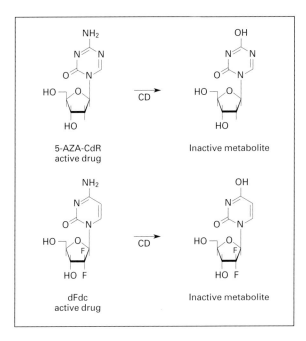

Fig. 3. Conversion of 5-AZA-CdR and dFdC to inactive metabolites by enzymatic deamination with cytidine deaminase.

of our work was to confer drug resistance against cytosine nucleoside analogs in normal cells, by transfer of the gene for human CD.

Preclinical Studies Using the CD Gene for Chemoprotection

Human CD

CD converts, by irreversible hydrolytic deamination, cytidine and deoxycytidine to uridine and deoxyuridine, respectively [63] (fig. 3). The principal tissue involved in deamination is the liver since it was found to contain the highest CD activity [63, 64]. High levels of CD enzyme activity are also present in human spleen, lung, kidney, intestinal mucosa [64], placenta [65] and mature granulocytes [66]. It was shown that CD levels rise in direct correlation with the differentiation status of normal and leukemic granulocytes [66]. Shröder et al. [67] noted that human CD34+ peripheral blood progenitors contain low CD activity. In accordance with these observations is the fact that the induction of differentiation of the human leukemic cell line HL-60 by 5-AZA-CdR produced a significant augmentation in CD activity [68].

The human CD, purified from placenta, was estimated to have a molecular weight of 48.7 kD and to contain several identical subunits of about 16 kD [65]. These findings are similar to those obtained following purification of this enzyme from human normal and leukemic granulocytes [66], leukemic myeloblasts [69] and placenta [70].

A cDNA clone for CD was isolated from a human liver cDNA library, sequenced and revealed to consist of 910 bp [65]. It contains a 441-bp coding region for a 146-amino-acid protein. Other investigators have also isolated the cDNA for the human CD [71–73]. The chromosomal locus of the human CD gene was determined to be 1p35-36.2 [74].

In addition to the natural substrates, cytidine and deoxycytidine, CD also catalyzes the deamination of cytosine nucleoside analogs including the antineoplastic agents ARA-C, dFdC and 5-AZA-CdR causing a loss of their pharmacologic activity [9, 10, 63]. CD may be involved in drug resistance to cytosine nucleoside analogs since elevated enzyme activity in leukemic cells was reported for some patients at relapse after chemotherapy with these agents [75, 76]. The high CD levels present in human liver explains the short half-life of these analogs [64].

To increase the half-life of cytosine nucleoside analogs due to their rapid in vivo deamination, inhibitors of CD were initially investigated by Camiener and Smith [63]. The first potent inhibitor of CD was reported to be 3,4,5,6-tetrahydrouridine (THU) [77]. The inhibition produced by THU was identified as competitive and reversible [66, 78]. Kreis et al. [79] noted in patients with solid tumors, a considerable rise in plasma ARA-C levels subsequent to the administration of ARA-C in combination with THU. Other inhibitors of CD have been synthesized, some of which have demonstrated higher potency than THU [80, 81].

The initial goals of our in vitro studies were to transduce mouse fibroblast and primary marrow cells with retroviral particles containing the CD cDNA and determine if this confers higher CD expression and drug resistance to the cytosine nucleoside analogs ARA-C, dFdC, and 5-AZA-CdR. Our next objective was to transplant mice with CD-transduced murine marrow cells and subsequent to treatment with ARA-C, determine in recipient mice if the CD proviral DNA can be detected and expressed in different tissues long after transplantation.

In vitro Studies Using the CD Gene

To determine the possible use of CD for gene transfer experiments, we first constructed the retroviral vector by insertion of the human CD cDNA into it [82]. This vector, called pMFG-CD (fig. 4), was used to transfect a murine fibroblast ecotropic packaging cell line. Clones of MFG-CD virus-

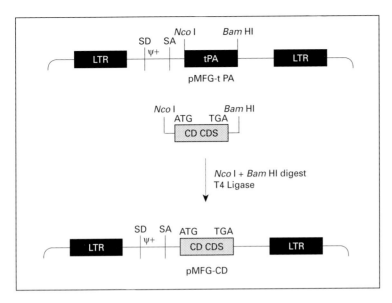

Fig. 4. Molecular design of the retroviral vector for the expression of CD cDNA (CD CDS) in target cells. pMFG-CD was constructed by cloning CD CDS between the *Nco* I and *Bam* HI sites. Ψ^+ = Packaging sequence; SD = splice donor; SA = splice acceptor site [82].

producing cells were isolated and demonstrated to have an elevated CD enzyme activity, increased expression of CD mRNA and protein, and to possess the ARA-C resistance phenotype.

Using the MFG-CD virions generated by one producer cell line, we transduced NIH 3T3 mouse fibroblasts [83]. Two clones of transduced cells were isolated after ARA-C selection, 3T3-CD3-V5 (V5) and 3T3-CD3-V6 (V6), and shown to overexpress the CD transgene. These clones also displayed resistance to the inhibitory action of ARA-C on cell growth, on DNA synthesis, and on colony formation. Clone V5 exhibited a higher CD enzyme activity and more resistance to ARA-C as compared to clone V6. This result indicates that there is a correlation between the extent of expression of the CD transgene and drug resistance to ARA-C. Furthermore, clone V5 showed stronger expression of the spliced proviral mRNA transcript suggesting that the higher splicing correlated with increased CD enzyme activity and increased ARA-C resistance. Our results are in agreement with those of Krall et al. [84] who determined in hematopoietic cells, using also the MFG vector, that enhanced amounts of spliced proviral mRNA correlated with higher transgene expression. In addition, we observed that by the use of continuous exposure to increasing concen-

Table 1. The antineoplastic action of ARA-C, 5-AZA-CdR, and dFdC on CD-transduced 3T3-CD3-V5 cells and effect of THU

Agent added	Loss of clonogenicity, %	
	3T3 (control)	3T3-CD3-V5 (transduced)
ARA-C (10^{-6} M)	92	5
ARA-C (10^{-6} M) + THU (10^{-4} M)	92	90
5-AZA-CdR (10^{-7} M)	82	20
5-AZA-CdR (10^{-7} M) + THU (10^{-4} M)	84	72
dFdC (10^{-7} M)	99	8
dFdC (10^{-7} M) + THU (10^{-4} M)	97	99

Determined by colony assay for a 50 h drug exposure. With THU alone, loss of clonogenicity was <5%. Modified from Eliopoulos et al. [86].

trations of ARA-C, it was possible to considerably enhance the expression of the CD transgene and drug resistance in the transduced cells [85].

Our next goal was to investigate whether CD gene transfer could confer drug resistance to other cytosine nucleoside analogs, such as the promising experimental drugs, 5-AZA-CdR and dFdC [86]. We analyzed the V5 clone by clonogenic assay and observed that these cells were resistant not only to ARA-C, but were also cross-resistant to 5-AZA-CdR and dFdC (table 1). In order to demonstrate that the drug resistance phenotype was due to the overexpression of the CD gene, we used THU, a competitive inhibitor of the CD enzyme. We showed that THU restored the sensitivity of the V5 clone to the inhibition of colony formation produced by cytosine nucleoside analogs (table 1). Our following aim was to determine whether the reversal of drug resistance in CD-transduced cells by THU correlated with its inhibition of CD enzyme activity. To accomplish this, we measured the CD activity in cytosolic extracts of these cells, in the presence or absence of THU. We showed that there was a high correlation between the inhibition of CD activity produced by THU and its reversal of drug resistance. These experiments using THU support that the major cause of the drug resistance to cytosine nucleoside analogs is the augmented CD activity in the CD-transduced cells. An interesting application of our CD-transduced cells would be as a model for evaluating the potency of new inhibitors of the CD enzyme activity. Neff and Blau [87] also showed in vitro chemoresistance to ARA-C following gene transfer of CD into 3T3 cells and lymphocytes using a different retroviral construct, the LCDSN vector. They demonstrated that THU could reverse the ARA-C resistance in CD-transduced lymphocytes. Another group of investigators, Schröder et al. [67], introduced the cDNA for CD into A9 murine fibroblast cells and showed an increase in CD enzyme activity and a reduction in the inhibition of growth and DNA synthesis produced by ARA-C.

In order to evaluate the potential of the CD transgene for hematopoietic chemoprotection, we transduced murine marrow cells with the MFG-CD retroviral vector [83]. We obtained an 80% efficiency of gene transfer determined by PCR analysis of colonies for the presence of the CD transgene. Clonogenic assay on the CD-transduced marrow cells revealed significant resistance to ARA-C. The hematopoietic colonies from the CD-transduced cells demonstrated almost complete drug resistance to even the highest concentrations ($10^{-6}\,M$) of ARA-C.

In vitro Studies Using a Bicistronic Vector Containing the CD Gene

The use of bicistronic gene vectors to express two different genes was initially developed by Jang et al. [88]. These investigators showed that initiation of translation of some genes in the encephalomyocarditis virus results from the binding of ribosomes to an internal segment of the 5'-nontranslated region (5' NTR) [88]. A short section of the 5' NTR (only 574 bp), designated the internal ribosome entry site (IRES), was demonstrated to be necessary to allow efficient binding of ribosomes.

The bicistronic vectors may be useful in the gene therapy of cancer. Since most standard clinical regimens utilize combinations of anticancer agents in order to increase clinical effectiveness, it is of interest to develop vectors which allow the expression of two different classes of drug resistance genes. Bicistronic vectors have the advantage of permitting translation of two genes from one mRNA transcript driven by a single promoter and therefore avoiding dual promoter interaction [89]. Bicistronic vectors expressing two drug resistance genes have already been shown to function in vitro; for example, the combination of the genes MDR with MGMT [90] and the genes MDR with DHFR [91].

For our in vitro studies we have developed a bicistronic vector capable of expressing efficiently the human mutated DHFR (P31S) and the human CD gene [92]. In our retroviral construct, we placed the DHFR gene under the control of the MoMLV promoter (LTR) and the CD gene downstream of the IRES element (fig. 5). This combination of drug resistance genes can confer resistance to both antifolates and cytosine nucleoside analog drugs.

Gene expression from an IRES-containing construct is highly dependent on its structural integrity as well as on the length of the cDNA cloned upstream or downstream to the IRES. The shorter the cDNA, the lower the probability of producing structural interference with the IRES element. The cDNA coding

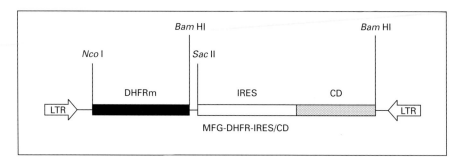

Fig. 5. Retroviral MFG-DHFR-IRES/CD vector structure. This bicistronic vector contains upstream from the IRES element the DHFR cDNA mutated (DHFRm) at codon 31 a phenylalanine was replaced by serine and downstream the CD cDNA [92].

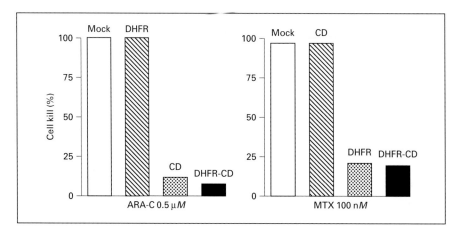

Fig. 6. Survival of transduced CFU-C at concentrations of ARA-C (*a*) or MTX (*b*). Mock, MFG-DHFR, MFG-CD and MFG-DHFR-IRES/CD (DHFR-CD) transduced murine bone marrow cells were plated in 1% methylcellulose. Colonies were counted and percent cell kill determined [92].

regions of both DHFR and CD are relatively small, 564 and 441 bp, respectively. This represents a marked advantage when compared to larger cDNAs and should favor efficient translation of both genes with the IRES element.

The effectiveness of chemoprotection with our bicistronic vector construct MFG-DHFR-IRES/CD was evaluated in vitro in murine bone marrow cells (fig. 6) [92]. We were interested in comparing its activity to the single gene vectors MFG-CD and MFG-DHFR bearing only one drug resistance gene. Using high titer retroviral producer cells, we were able to transduce over 90% of the clonogenic progenitor cells. We observed a high level of drug resistance that was comparable to that obtained with the single gene vectors. The extent of drug resistance to cell kill produced by ARA-C was similar in the cells transduced with the MFG-CD vector as compared to the cells transduced with the bicistronic vector (fig. 6a). Conversely, the extent of drug resistance to cell kill produced by methotrexate (MTX) was almost identical for the DHFR-transduced cells as compared to the DHFR-CD-transduced cells (fig. 6b). These results indicate that bicistronic vectors containing the CD gene can function efficiently to confer drug resistance to two different classes of agents.

In vivo Studies with the CD Gene in Hematopoietic Cells

Our next goal was to determine whether it was possible to obtain long-term in vivo transgene expression and drug resistance following transplantation in mice of murine hematopoietic cells transduced with MFG-CD [93]. In these experiments, the CD-transduced primary bone marrow cells from male mice were transplanted into female syngeneic mice. All mice received three cycles of ARA-C at intervals of greater than 6 weeks. At over 10 months after transplantation, the CD proviral DNA was detected by PCR in the majority of blood, marrow and spleen samples obtained from the recipient mice, demonstrating long-term persistence of the CD proviral DNA sequences. These findings suggest that the transduction of pluripotent hematopoietic cells possessing the capacity of long-term engraftment occurred in these experiments.

We also conducted clonogenic assays and determined by PCR analysis that many of the hematopoietic colonies formed in vitro from recipient mice were positive for the MFG-CD provirus. The presence of the CD transgene in these colonies was detected more than 10 months from transplantation, signifying that the marrow progenitors that had been successfully transduced were capable of surviving in mice and capable of producing in vitro colonies. In addition, in some recipient mice, we detected signs of in vitro drug resistance to ARA-C in the hematopoietic cells harvested 11–13 months after transplantation (table 2).

Assays for CD enzyme activity of hematopoietic tissues obtained from some recipient CD mice at 11–13 months following transplantation, displayed significantly higher levels of activity than the control group, recipient Lac mice (table 2). These results demonstrate long-term CD transgene in vivo expression which can persist for at least 11 months. In some mice, the low CD enzyme activity detected in different tissues assayed after transplantation may have resulted from low efficiency of transduction, low donor cell engraftment, and/or inactivation of CD transgene expression. Low expression of the target gene in the MFG retroviral vector has been reported to sometimes occur in vivo and be related to de novo methylation of cytosine residues within

Table 2. CD enzyme activity and in vitro ARA-C resistance in hematopoietic cells of mice transplanted with CD-transduced marrow cells

Mouse	CD enzyme activity[1] in marrow (U/mg)	Survival (%) 0.5 µM ARA-C[2]	
		Marrow	Spleen
CD#3	3.1	6	ND
CD#4	2.5	2	23
CD#6	49.5	29	17
Lac mice	0.6	<1	<1

Mouse CD#: mouse transplanted with CD-transduced marrow cells. Modified from Eliopoulos et al. [93].

[1] Unit: deamination of 1 nmol cytidine per min.
[2] Determined by colony assay. ND = Not determined.

the MoMLV LTR promoter region [94]. Notably, recipient mouse CD#6 exhibited a very large increase in CD enzyme activity in marrow, spleen, and blood cells [93]. This mouse also showed the highest in vitro drug resistance. One possible explanation for these results is that the in vivo ARA-C treatments may have selected for high CD-expressing hematopoietic cells.

Conclusions and Perspectives

Our in vitro data indicate that it is possible to confer drug resistance to cytosine nucleoside analogs by transfer of the CD gene into murine hematopoietic cells. Our study in mice is the first demonstration of in vivo CD transgene expression which shows that it is possible to transplant CD-transduced marrow cells and obtain long-term increased expression of CD in hematopoietic cells [93]. These in vivo data hold promise for the use of the CD gene for chemoprotection to improve the therapeutic efficacy of cytosine nucleoside analogs in cancer treatment.

Since cytosine nucleoside analogs, such as ARA-C, dFdC and 5-AZA-CdR, are S-phase-specific agents, their major dose-limiting toxicity is myelosuppression. The doses of these drugs can be increased considerably without encountering serious nonhematopoietic toxicity. The major advantage of intensive doses in tumor therapy is that more cytotoxic drug can penetrate the central core of the target tumor. This is essential to achieve curative therapy or long-term survival. Both dFdC and 5-AZA-CdR show promising activity in tumor therapy, especially in non-small cell lung cancer [7, 8] where conventional chemotherapy has little or no success in the treatment of the metastatic form of this disease. ARA-C also produced an interesting response in advanced breast cancer that should be further investigated [6].

Novel approaches for the chemotherapy of advanced malignant diseases should be explored. The use of gene therapy for chemoprotection against the hematopoietic toxicity produced by cytosine nucleoside analogs is one approach that may have enormous potential in the chemotherapy of metastatic cancer which is not very responsive to conventional chemotherapy.

The success of gene therapy for protecting normal cells from the toxicity of cytosine nucleoside analogs depends not only on the functional state of the CD gene, but also on the type of vector and packaging cell line used. Progress made in these areas will help promote the use of CD for chemoprotection. The utilization of better gene delivery systems and improved packaging cell lines that generate very high viral titers and viral particles of greater stability will enhance the transfer and expression of the CD gene in target hematopoietic cells. One advantage of using CD for chemoprotection is that it is a small gene (441 bp) which facilitates genetic manipulation in design of new vectors for this purpose.

An aspect that may facilitate the use of chemoprotection is that it is only required during drug treatment. Furthermore, the design of bicistronic vectors containing CD and an additional drug resistance gene may be important tools to improve therapeutic efficacy by conferring protection against a second drug which produces a synergistic antineoplastic effect in combination with the cytosine nucleoside analog.

Animal models should be employed to optimize the conditions for chemoprotection with the CD gene and to test its application against tumors. An important advancement would be to demonstrate that chemotherapy with cytosine nucleoside analogs is more effective in the animals transplanted with CD-transduced hematopoietic cells, as compared to the control mice. The final step in establishing the therapeutic effectiveness of chemoprotection using the CD gene will be a clinical study in patients with advanced cancer who will be administered intensive treatments with cytosine nucleoside analogs.

Acknowledgement

This work was supported by grant MT-13754 from the Medical Research Council of Canada.

References

1 Wingo PA, Tong T, Bolden S: Cancer Statistics, 1995. Cancer J Clinicians 1995;45:8–30.
2 Corey CA, DeSilva AD, Holland CA, Williams DA: Serial transplantation of methotrexate-resistant bone marrow: Protection of murine recipients from drug toxicity by progeny of transduced stem cells. Blood 1990;75:337–343.
3 Hanania EG, Fu S, Roninson I, Zu Z, Gottesman MM, Deisseroth AB: Resistance to taxol chemotherapy produced in mouse marrow cells by safety-modified retroviruses containing a human MDR-1 transcription unit. Gene Ther 1995;2:279–284.
4 Li MX, Banerjee D, Zhao SC, Schweitzer BI, Mineishi S, Gilboa E, Bertino JR: Development of a retroviral construct containing a human mutated dihydrofolate reductase cDNA for hematopoietic stem cell transduction. Blood 1994;83:3403–3408.
5 Podda S, Ward M, Himelstein A, Richardson C, de la Flor-Weiss E, Smith L, Gottesman M, Pastan I, Bank A: Transfer and expression of the human multiple drug resistance gene into live mice. Proc Natl Acad Sci USA 1992;89:9676–9680.
6 Czaykowski PM, Samuels T, Oza A: A durable response to cytarabine in advanced breast cancer. Clin Oncol 1997;9:181–183.
7 Fossella FV, Lippman SM, Shin DM, Tarassoff P, Calayag-Jung M, Perez-Soler R, Lee JS, Murphy WK, Glisson B, Rivera E, Hong WK: Maximum-tolerated dose defined for single-agent gemcitabine: A phase I dose-escalation study in chemotherapy-naive patients with advanced non-small-cell lung cancer. J Clin Oncol 1997;15:310–316.
8 Momparler RL, Bouffard DY, Momparler LF, Dionne J, Bélanger K, Ayoub J: Pilot phase I-II study on 5-aza-2′-deoxycytidine (Decitabine) in patients with metastatic lung cancer. Anti-Cancer Drugs 1997;8:358–368.
9 Bouffard DY, Laliberté J, Momparler RL: Kinetic studies on 2′,2′-difluorodeoxycytidine (gemcitabine) with purified human deoxycytidine kinase and cytidine deaminase. Biochem Pharmacol 1993; 45:1857–1861.
10 Chabot GG, Bouchard J, Momparler RL: Kinetics of deamination of 5-aza-2′-deoxycytidine and cytosine arabinoside by human liver cytidine deaminase and its inhibition by 3-deazauridine, thymidine or uracil arabinoside. Biochem Pharmacol 1983;32:1327–1328.
11 Grant S: Ara-C: Cellular and molecular pharmacology. Adv Cancer Res 1998;72:197–233.
12 Momparler RL, Fischer GA: Mammalian deoxynucleoside kinases. 1. Deoxycytidine kinase: Purification. Properties and kinetic studies with cytosine arabinoside. J Biol Chem 1968;243:4298–4304.
13 Momparler RL: Effect of cytosine arabinoside 5′-triphosphate on mammalian DNA polymerase. Biochem Biophys Res Commun 1969;34:464–471.
14 Karon M, Shirakawa S: The locus of action of 1-beta-D-arabinofuranosylcytosine in the cell cycle. Cancer Res 1969;29:687–696.
15 Keating MJ, McCredie KB, Bodey GP, Smith TL, Gehan E, Freireich EJ: Improved prospects for long-term survival in adults with acute myelogenous leukemia. J Am Med Assoc 1982;248: 2481–2486.
16 Stryckmans P, DeWitte T, Bitar N, Marie JP, Suciu S, Sohbu G, Debusscher L, Bury J, Peetermans M, Andrien JM, Fiere D, Bron D, Dekker A, Zittoun R: Cytosine arabinoside for induction, salvage, and consolidation therapy of adult acute lymphoblastic leukemia. Semin Oncol 1987;14: 67–72.
17 Kantarjian H, Barlogie B, Plunkett W, Velasquez W, McLaughlin P, Riggs S, Cabanillas F: High-dose cytosine arabinoside in non-Hodgkin's lymphoma. J Clin Oncol 1983;1:689–694.
18 Shipp MA, Takvorian RC, Canellos GP: High-dose cytosine arabinoside. Active agent in treatment of non-Hodgkin's lymphoma. Am J Med 1984;77:845–850.
19 Kern DH, Morgan CR, Hildebrand-Zanki SU: In vitro pharmacodynamics of 1-beta-D-arabinofuranosylcytosine: Synergy of antitumor activity with cis-diamminedichloroplatinum (II). Cancer Res 1988;48:117–121.
20 Chabner BA: Cytidine analogues; in Chabner BA, Longo DL (eds): Cancer Chemotherapy and Biotherapy, ed 2. Philadelphia, Lippincott-Raven, 1996, pp 213–233.

21 Parkinson DR, Pluda JM, Cazenave L, Ho P, Sorensen JM, Sznol M, Christian MC: Investigational anticancer agents; in Chabner BA, Longo DL (eds): Cancer Chemotherapy and Biotherapy, ed 2. Philadelphia, Lippincott-Raven, 1996, pp 509–528.
22 Plunkett W, Huang P, Gandhi V: Preclinical characteristics of gemcitabine. Anti-Cancer Drugs 1995;6(suppl 6):7–13.
23 Huang P, Chubb S, Hertel LW, Grindey GB, Plunkett W: Action of 2′,2′-difluorodeoxycytidine on DNA synthesis. Cancer Res 1991;51:6110–6117.
24 Noble S, Goa KL: Gemcitabine. A review of its pharmacology and clinical potential in non-small cell lung cancer and pancreatic cancer. Drugs 1997;54:447–472.
25 Carmichael J, Walling J: Advanced breast cancer: Investigational role of gemcitabine. Eur J Cancer 1997;33(suppl 1):S27–S30.
26 Burris HA, Moore MJ, Andersen J, Green MR, Rothenberg ML, Modiano MR, Cripps MC, Portenoy RK, Storniolo AM, Tarassoff P, Nelson R, Dorr FA, Stephens CD, Von Hoff DD: Improvements in survival and clinical benefit with gemcitabine as first-line therapy for patients with advanced pancreas cancer: A randomized trial. J Clin Oncol 1997;15:2403–2413.
27 Casper ES, Green MR, Kelsen DP, Heelan RT, Brown TD, Flombaum CD, Trochanowski B, Tarassoff PG: Phase II trial of gemcitabine (2′,2′-difluorodeoxycytidine) in patients with adenocarcinoma of the pancreas. Invest New Drugs 1994;12:29–34.
28 Stadler WM, Kuzel TM, Raghavan D, Levine E, Vogelzang NJ, Roth B, Dorr FA: Metastatic bladder cancer: Advances in treatment. Eur J Cancer 1997;33(suppl 1):S23–S26.
29 Kaufmann M, von Minckwitz G: Gemcitabine in ovarian cancer: An overview of safety and efficacy. Eur J Cancer 1997;33:S31–S33.
30 Lund B, Hansen OP, Theilade K, Hansen M, Neijt JP: Phase II study of gemcitabine (2′,2′-difluorodeoxycytidine) in previously treated ovarian cancer patients. J Natl Cancer Inst 1994;86: 1530–1533.
31 Abratt RP, Bezwoda WR, Falkson G, Goedhals L, Hacking D, Rugg TA: Efficacy and safety profile of gemcitabine in non-small-cell lung cancer: A phase II study. J Clin Concol 1994;12:1535–1540.
32 Cormier Y, Eisenhauer E, Muldal A, Gregg R, Ayoub J, Goss G, Stewart D, Tarasoff P, Wong D: Gemcitabine is an active new agent in previously untreated extensive small cell lung cancer (SCLC). A study of the National Cancer Institute of Canada Clinical Trials Group. Ann Oncol 1994;5: 283–285.
33 Catimel G, Vermorken JB, Clavel M, De Mulder P, Judson I, Sessa C, Piccart M, Bruntsch U, Verweij J, Wanders J, Franklin H, Kaye SB: A phase II study of Gemcitabine (LY188011) in patients with advanced squamous cell carcinoma of the head and neck. Ann Oncol 1994;5:543–547.
34 Momparler RL: Molecular, cellular and animal pharmacology of 5-aza-2′-deoxycytidine. Pharmacol Therapeut 1985;30:287–299.
35 Jones PA, Taylor SM: Cellular differentiation, cytidine analogs and DNA methylation. Cell 1980; 20:85–93.
36 Jones PA: DNA methylation errors and cancer. Cancer Res 1996;56:2463–2467.
37 Razin A, Cedar H: DNA methylation and gene expression. Microbiol Rev 1991;55:451–458.
38 Momparler RL, Bouchard J, Samson J: Induction of differentiation and inhibition of DNA methylation in HL-60 myeloid leukemic cells by 5-AZA-2′-deoxycytidine. Leuk Res 1985;9:1361–1366.
39 Bender CM, Pao MM, Jones PA: Inhibition of DNA methylation by 5-aza-2′-deoxycytidine suppresses the growth of human tumor cell lines. Cancer Res 1998;58:95–101.
40 Momparler RL, Rivard GE, Gyger M: Clinical trial on 5-aza-2′-deoxycytidine in patients with acute leukemia. Pharmacol Ther 1985;30:277–286.
41 Richel DJ, Colly LP, Kluin-Nelemans JC, Willemze R: The antileukaemic activity of 5-aza-2′-deoxycytidine (Aza-dC) in patients with relapsed and resistant leukaemia. Br J Cancer 1991;64: 144–148.
42 Rivard GE, Momparler RL, Demers J, Benoit P, Raymond R, Lin KT, Momparler LF: Phase I study on 5-aza-2′-deoxycytidine in children with acute leukemia. Leuk Res 1981;5:453–462.
43 Zagonel V, Lo Re G, Marotta G, Babare R, Sardeo G, Gattei V, De Angelis V, Monfardini S, Pinto A: 5-Aza-2′-deoxycytidine (Decitabine) induces trilineage response in unfavourable myelodysplastic syndromes. Leukemia 1993;7(suppl 1):30–35.

44　Herman JG, Latif F, Weng Y, Lerman MI, Zbar B, Liu S, Samid D, Duan DS, Gnarra JR, Linehan WM, Baylin SB: Silencing of the VHL tumor-suppressor gene by DNA methylation in renal carcinoma. Proc Natl Acad Sci USA 1994;91:9700–9704.

45　Côté S, Momparler RL: Activation of the retinoic acid receptor beta gene by 5-aza-2'-deoxycytidine in human DLD-1 colon carcinoma cells. Anti-Cancer Drugs 1997;8:56–61.

46　Côté S, Sinnett D, Momparler RL: Demethylation by 5'-aza-2'-deoxycytidine of specific 5-methylcytosine sites in the promoter region of the retinoic acid receptor beta gene in human colon carcinoma cells. Anti-Cancer Drugs 1998;9:743–750.

47　Merlo A, Herman JG, Mao L, Lee DJ, Gabrielson E, Burger PC, Baylin SB, Sidransky D: 5'CpG island methylation is associated with transcriptional silencing of the tumor suppressor p16/CDKN2/MTS1 in human cancers. Nat Med 1995;1:686–692.

48　Otterson GA, Khleif SN, Chen W, Coxon AB, Kaye FJ: CDKN2 gene silencing in lung cancer by DNA hypermethylation and kinetics of p16 INK4 protein induction by 5-aza-2'-deoxycytidine. Oncogene 1995;11:1211–1216.

49　Costello JF, Berger MS, Huang HS, Cavenee WK: Silencing of p16/CDKN2 expression in human gliomas by methylation and chromatin condensation. Cancer Res 1996;56:2405–2410.

50　Gonzalgo ML, Hayashida T, Bender CM, Pao MM, Tsai YC, Gonzales FA, Nguyen HD, Nguyen TT, Jones PA: The role of DNA methylation in expression of the p19/p16 locus in human bladder cancer cell lines. Cancer Res 1998;58:1245–1252.

51　Graff JR, Herman JG, Lapidus RG, Chopra H, Xu R, Jarrard DF, Isaacs WB, Pitha PM, Davidson NE, Baylin SB: E-cadherin expression is silenced by DNA hypermethylation in human breast and prostate carcinomas. Cancer Res 1995;55:5195–5199.

52　Bodensteiner DC, Doolittle GC: Adverse haematological complications of anticancer drugs. Clinical presentation, management and avoidance. Drug Safety 1993;8:213–224.

53　Eddy DM: High-dose chemotherapy with autologous bone marrow transplantation for the treatment of metastatic breast cancer. J Clin Oncol 1992;10:657–670.

54　Bertino JR: 'Turning the tables' – Making normal marrow resistant to chemotherapy. J Natl Cancer Inst 1990;82:1234–1235.

55　Banerjee D, Zhao SC, Li MX, Schweitzer Bl, Mineishi S, Bertino JR: Gene therapy utilizing drug resistance genes: A review. Stem Cells 1994;12:378–385.

56　Koç ON, Allay JA, Keunmyoung L, Davis BM, Reese JS, Gerson SL: Transfer of drug resistance genes into hematopoietic progenitors to improve chemotherapy tolerance. Semin Oncol 1996;23:46–65.

57　Rafferty JA, Hickson I, Chinnasamy N, Lashford LS, Margison GP, Dexter TM, Fairbairn LJ: Chemoprotection of normal tissues by transfer of drug resistance genes. Cancer Metast Rev 1996;15:365–383.

58　Zhao SC, Banerjee D, Mineishi S, Bertino JR: Post-transplant methotrexate administration leads to improved curability of mice bearing a mammary tumor transplanted with marrow transduced with a mutant human dihydrofolate reductase cDNA. Hum Gene Ther 1997;8:903–909.

59　Deisseroth AB, Kavanagh J, Champlin R: Use of safety-modified retroviruses to introduce chemotherapy resistance sequences into normal hematopoietic cells for chemoprotection during the therapy of ovarian cancer: A pilot trial. Hum Gene Ther 1994;5:1507–1522.

60　Deisseroth AB, Holmes F, Hortobagyi G, Champlin R: Use of safety-modified retroviruses to introduce chemotherapy resistance sequences into normal hematopoietic cells for chemoprotection during the therapy of breast cancer: A pilot trial. Hum Gene Ther 1996;7:401–416.

61　O'Shaughnessy JA: Chemoprevention of breast cancer. JAMA 1996;275:1349–1353.

62　O'Shaughnessy JA, Cowan KH, Nienhuis AW, McDonagh KT, Sorrentino BP, Dunbar CE, Chiang Y, Wilson W, Goldspiel B, Kohler D, Cottler-Fox M, Leitman S, Gottesman M, Pastan I, Denicoff A, Noone M, Gress R: Retroviral mediated transfer of the human multidrug resistance gene (MDR-1) into hematopoietic stem cells during autologous transplantation after intensive chemotherapy for metastatic breast cancer. Hum Gene Ther 1994;5:891–911.

63　Camiener GW, Smith CG: Studies of the enzymatic deamination of cytosine arabinoside. I. Enzyme distribution and species specificity. Biochem Pharmacol 1965;14:1405–1416.

64　Ho DHW: Distribution of kinase and deaminase of 1β-D-arabinofuranosylcytosine in tissues of man and mouse. Cancer Res 1973;33:2816–2820.

65   Laliberté J, Momparler RL: Human cytidine deaminase: Purification of enzyme, cloning, and expression of its complementary DNA. Cancer Res 1994;54:5401–5407.
66   Chabner BA, Johns DG, Coleman CN, Drake JC, Evans WH: Purification and properties of cytidine deaminase from normal and leukemic granulocytes. J Clin Invest 1974;53:922–931.
67   Schröder JK, Kirch C, Flasshove M, Kalweit H, Seidelmann M, Hilger R, Seeber S, Schütte J: Constitutive overexpression of the cytidine deaminase gene cofers resistance to cytosine arabinoside in vitro. Leukemia 1996;10:1919–1924.
68   Momparler RL, Laliberté J: Induction of cytidine deaminase in HL-60 myeloid leukemic cells by 5-aza-2′-deoxycytidine. Leuk Res 1990;14:751–754.
69   Cheng YC, Tan RS, Ruth JL, Dutschman G: Cytotoxicity of 2′-fluoro-5-iodo-1β-D-arabinofuranosyl-cytosine and its relationship to deoxycytidine deaminase. Biochem Pharmacol 1983;32:726–729.
70   Cacciamani T, Vita A, Cristalli G, Vincenzetti S, Natalini P, Ruggieri S, Amici A, Magni G: Purification of human cytidine deaminase: Molecular and enzymatic characterization and inhibition by synthetic pyrimidine analogs. Arch Biochem Biophys 1991;290:285–292.
71   Kühn K, Bertling WM, Emmrich F: Cloning of a functional cDNA for human cytidine deaminase (CDD) and its use as a marker of monocyte/macrophage differentiation. Biochem Biophys Res Commun 1993;190:1–7.
72   Vincenzetti S, Cambi A, Neuhard J, Garattini E, Vita A: Recombinant human cytidine deaminase: Expression, purification, and characterization. Protein Express Purif 1996;8:247–253.
73   Gran C, Bøyum A, Johansen RF, Løvhaug D, Seeberg EC: Growth inhibition of granulocyte-macrophage colony-forming cells by human cytidine deaminase requires the catalytic function of the protein. Blood 1998;91:4127–4135.
74   Saccone S, Besati C, Andreozzi L, Della Valle G, Garattini E, Terao M: Assignment of the human cytidine deaminase (CDA) gene to chromosome 1 band p35-p36.2. Genomics 1994;22:661–662.
75   Onetto N, Momparler RL, Momparler LF, Gyger M: In vitro biochemical tests to evaluate the response to therapy of acute leukemia with cytosine arabinoside or 5-aza-2′-deoxycytidine. Semin Oncol 1987;14:231–237.
76   Steuart CD, Burke PJ: Cytidine deaminase and the development of resistance to arabinosyl cytosine. Nat New Biol 1971;233:109–110.
77   Camiener GW: Studies of the enzymatic deamination of aracytidine. V. Inhibition in vitro and in vivo by tetrahydrouridine and other related pyrimidine nucleosides. Biochem Pharmacol 1968;17:1981–1991.
78   Wentworth DF, Wolfenden R: On the interaction of 3,4,5,6-tetrahydrouridine with human liver cytidine deaminase. Biochemistry 1975;14:5099–5105.
79   Kreis W, Chan K, Budman DR, Schulman P, Allen S, Weiselberg L, Lichtman S, Henderson V, Freeman J, Deere M, Andreeff M, Vinciguerra V: Effect of tetrahydrouridine on the clinical pharmacology of 1-beta-D-arabinofuranosylcytosine when both drugs are coinfused over three hours. Cancer Res 1988;48:1337–1342.
80   Marquez VE, Liu PS, Kelley JA, Driscoll JS, McCormack JJ: Synthesis of 1,3-diazepin-2-one nucleosides as transition-state inhibitors of cytidine deaminase. J Med Chem 1980;23:713–775.
81   Laliberté J, Marquez VE, Momparler RL: Potent inhibitors for the deamination of cytosine arabinoside and 5-aza-2′-deoxycytidine by human cytidine deaminase. Cancer Chemother Pharmacol 1992;30:7–11.
82   Momparler RL, Laliberté J, Eliopoulos N, Beauséjour C, Cournoyer D: Transfection of murine fibroblast cells with human cytidine deaminase cDNA confers resistance to cytosine arabinoside. Anti-Cancer Drugs 1996;7:266–274.
83   Momparler RL, Eliopoulos N, Bovenzi V, Létourneau S, Greenbaum M, Cournoyer D: Resistance to cytosine arabinoside by retroviral-mediated gene transfer of human cytidine deaminase into murine fibroblast and hematopoietic cells. Cancer Gene Ther 1996;3:331–338.
84   Krall WJ, Skelton DC, Yu XJ, Riviere I, Lehn P, Mulligan RC, Kohn DB: Increased levels of spliced RNA account for augmented expression from the MFG retroviral vector in hematopoietic cells. Gene Ther 1996;3:37–48.

85 Beauséjour CM, Momparler RL: Gene amplification of human cytidine deaminase proviral cDNA and increased levels of its mRNA produces enhanced drug resistance to cytosine arabinoside in retroviral-transduced murine fibroblasts. Cancer Lett, in press.

86 Eliopoulos N, Cournoyer D, Momparler RL: Drug resistance to 5-aza-2′-deoxycytidine, 2′,2′-difluorodeoxycytidine and cytosine arabinoside conferred by retroviral-mediated transfer of human cytidine deaminase cDNA into murine cells. Cancer Chemother Pharmacol 1998;42:373–378.

87 Neff T, Blau CA: Forced expression of cytidine deaminase confers resistance to cytosine arabinoside and gemcitabine. Exp Hematol 1996;24:1340–1346.

88 Jang SK, Krausslich H-G, Nicklin MJH, Duke GM, Palmenberg AC, Wimmer E: A segment of the 5′ nontranslated region of encephalomyocarditis virus RNA directs internal entry of ribosomes during in vitro translation. J Virol 1988;62:2636–2643.

89 Emerman M, Temin HH: Quantitative analysis of gene suppression in integrated retrovirus vectors. Mol Cell Biol 1986;6:792–800.

90 Suzuki M, Sugimoto Y, Tsukahara S, Okochi E, Gottesman MM, Tsuruo T: Retroviral coexpression of two different types of drug resistance genes to protect normal cells from combination chemotherapy. Clin Cancer Res 1997;3:947–954.

91 Galipeau J, Benaim E, Spencer HT, Blakeley RL, Sorrentino BP: A bicistronic retroviral vector for protecting hematopoietic cells against antifolates and p-glycoprotein effluxed drugs. Hum Gene Ther 1997;8:1773–1783.

92 Beauséjour CM, Le NLO, Létourneau S, Cournoyer D, Momparler RL: Coexpression of cytidine deaminase and mutant dihydrofolate reductase by a bicistronic retroviral vector confers resistance to cytosine arabinoside and methotrexate. Hum Gene Ther 1998;9:2537–2544.

93 Eliopoulos N, Bovenzi V, Le NLO, Momparler LF, Greenbaum M, Létourneau S, Cournoyer D, Momparler RL: Retroviral transfer and long-term expression of human cytidine deaminase cDNA in hematopoietic cells following transplantation in mice. Gene Ther 1988;5:1545–1551.

94 Challita PM, Kohn DB: Lack of expression from a retroviral vector after transduction of murine hematopoietic stem cells is associated with methylation in vivo. Proc Natl Acad Sci USA 1994;91:2567–2571.

Dr. Richard L. Momparler, Département de Pharmacologie, Université de Montréal,
Centre de Recherche Pédiatrique, Hôpital Ste-Justine,
3175 Côte Ste-Catherine, Montréal, Québec H3T 1C5 (Canada)
Tel. +1 514 345 4691, Fax +1 514 345 4801, E-Mail momparlr@ere.umontreal.ca

In vivo Selection of Hematopoietic Stem Cells Transduced with DHFR-Expressing Retroviral Vectors

B.P. Sorrentino[a], J.A. Allay[a], R.L. Blakley[b]

Departments of [a] Hematology and Oncology, and [b] Molecular Pharmacology, St. Jude Children's Research Hospital, Memphis, Tenn., USA

The Potential for Hematopoietic Stem Cell Gene Therapy

A wide variety of inherited genetic diseases are potentially curable by transplantation with normal hematopoietic stem cells. Diverse disorders such as the hemoglobinopathies, inherited immunodeficiency syndromes, and certain metabolic storage diseases are all caused by functional defects in stem cell progeny. One approach has been to use allogeneic stem cells from a normal individual to correct the patient's stem cell defect. Although considerable progress has been made using allogeneic transplantation to treat some of these disorders [1], significant toxicity has limited the scope and efficacy of this treatment modality. As a result, allogeneic transplantation has not been widely used to treat diseases such as sickle cell anemia, thalassemia, and chronic granulomatous disease. Histocompatibility differences between the donor and the patient constitute one of the major risks of allogeneic transplantation, and can result in life-threatening complications from graft versus host disease (GVHD) or graft rejection. Although limited GVHD can have a positive effect in treating certain malignancies, it is clearly a liability for the treatment of nonmalignant genetic diseases. Therapies that reverse or prevent GVHD are themselves toxic and can have detrimental effects on engraftment and host immunity.

One approach to avoid these limitations is the use of retroviral vectors to correct the patient's own stem cells [2]. These modified autologous cells can be reinfused without the risks of GVHD or graft failure, and may therefore be more suitable for the treatment of nonmalignant genetic diseases. Candidate diseases for stem cell gene therapy are increasing as the molecular etiologies for a growing number of hematopoietic disorders are defined. For instance, the recent discovery that the majority of severe combined immunodeficiency (SCID) cases are due to single gene defects in cytokine signaling has led to the development and testing of corrective genetic vectors [3–5]. Furthermore, recent studies have shown that gene replacement can be curative in a number of animal models of human disease [6–8]. Therefore, interest in stem cell gene therapy has been growing as the result of its potentially broad impact on a wide variety of human diseases [2].

The Possibility of Using in vivo Selection to Overcome Low Transduction Efficiencies

The main problem that currently thwarts hematopoietic stem cell gene therapy is the low number of genetically modified cells that are achieved after transplantation. Despite the fact that murine retroviral vectors can transduce a large proportion of reconstituting stem cells from mice [9–11], these vectors result in much lower proportions of modified cells in nonhuman primates [12] and in human marking trials [13, 14]. As a result of this problem, clinical gene therapy protocols have been limited by marking of less than 1% of the circulating peripheral blood cells [15, 16]. The biological explanations for this low transduction efficiency are just becoming elucidated, and include the lack of expression for key retroviral receptors on the surface of stem cells [17] and the block to viral integration associated with stem cell quiescence [18]. These impediments to efficient transduction may not be prohibitive for the treatment of certain disorders such as SCID, where the corrected cells have a naturally occurring selective advantage [8, 19]. In contrast, the refractoriness of stem cells to genetic modification is a major limitation for the treatment of most myeloid disorders.

One approach to this problem has been to develop new vectors and transduction conditions designed to overcome the biological limitations of stem cell transduction. An alternate but complementary approach is the development of a system for the enrichment and amplification of genetically modified stem cells, which could potentially compensate for the inefficiency of stem cell transduction associated with any vector system. The general strategy for stem cell selection centers on the incorporation of a dominant selectable marker into the therapeutic vector. Stem cells expressing such a vector are then enriched in vivo by treating the patient with a selective agent at some point after transplantation (fig. 1). The best characterized systems for in vivo selection employ a drug resistance gene as the selectable marker [20]. The

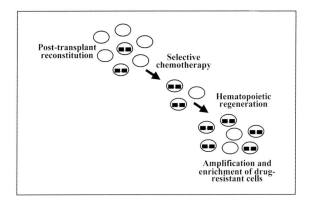

Fig. 1. In vivo selection of vector-transduced hematopoietic cells. Initially after transplant, only a small minority of engrafting stem cells contain the retroviral vector. However, if this vector contains a dominant delectable marker, in vivo selection can be accomplished. With this strategy, the transplanted animal is treated with a drug that confers a selective advantage to transduced cells. After this selective chemotherapy, enrichment in transduced stem cells occurs. With subsequent hematopoietic regeneration, an amplification and enrichment of drug-resistant cells is accomplished.

feasibility of this approach has been demonstrated in mouse models using the human multidrug resistance 1 (MDR1) gene [21, 22] and more recently with alkyltransferase genes [23, 24]. Although these particular systems have been shown to increase the proportion of marked cells within peripheral blood and bone marrow populations, it is not yet known whether they can induce selection at the level of stem cells, or whether they can be extrapolated for use in large animal models.

We have focused on the use of dihydrofolate reductase (DHFR) variants as dominant selectable markers for stem cell enrichment. These genes have a number of intrinsic advantages for stem cell selection. Point mutations in the active site of the enzyme confer very high levels of cellular resistance to antifolate drugs such as methotrexate (MTX) and trimetrexate (TMTX) [25, 26]. In addition, the relatively small size of the DHFR cDNA (690 bp) facilitates incorporation of a second linked gene into the vector construct. In terms of the drugs used for selection, antifolates are relatively safe for posttransplantation selection considering their extensive use in treating GVHD and their relative lack of long-term toxicity and DNA-damaging effects.

To test the feasibility of using DHFR-expressing retroviral vectors for stem cell selection, we have utilized a mouse transplantation model to test for stem cell selection. Mice were transplanted with bicistronic retroviral vectors expressing both a resistance-conferring DHFR cDNA (DHFR^{L22Y}) [27], and

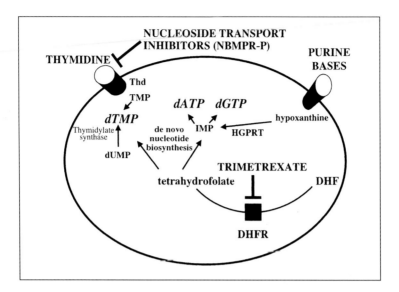

Fig. 2. Effects of TMTX and NBMPR on cellular nucleotide metabolism. TMTX inhibits DHFR function and thereby blocks de novo nucleotide biosynthesis. Hematopoietic stem cells can bypass this effect by salvaging thymidine and purine bases from the serum. NBMPR is a nucleoside transport inhibitor that potently blocks thymidine uptake in hematopoietic cells, thereby sensitizing them to the effects of TMTX. Reprinted from Allay et al. [34].

either a human CD24 [28] or green fluorescent protein (GFP) reporter gene [29]. This bicistronic design allows serial tracking of the number of vector-expressing cells within individual mice, both before and after drug treatment. Our results using this system for in vivo stem cell selection are summarized in this chapter.

Nucleotide Transport Inhibitors Increase Selective Pressure for Stem Cell Selection

Our intitial efforts were focused on deriving a drug treatment schedule suitable for stem cell selection. High doses of TMTX were given to normal mice over a 5-day treatment course that resulted in significant myelosuppression. Despite the toxic effects on peripheral blood counts, there was no decrease in the number of myeloid progenitors within the bone marrow. This selective sparing of early myeloid cells was initially thought to be due to cell cycle quiescence during the treatment interval, given that TMTX is an S-phase-

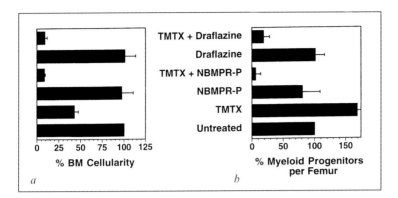

Fig. 3. Effects of nucleoside transport inhibitors on the content of myeloid progenitors in the femurs of drug-treated mice. Mice were treated with various combinations of TMTX and the nucleoside transport inhibitors NBMPR-P or draflazine for 5 days. On day 6, the bone marrow cellularity (*a*) and the progenitor content (*b*) was assayed. A 5-day course of TMTX that caused moderate bone marrow hypocellularity actually resulted in an increase in the number of progenitors. While neither transport inhibitor had an effect on hematopoiesis when used alone, both draflazine and NBMPR-P caused a marked sensitization of myeloid progenitors to the toxic effects of TMTX. Reprinted from Allay et al. [32].

specific drug. However, work from another group showed that when cycling was induced by treating mice with pegylated stem cell factor (SCF), both stem cells and myeloid progenitors remained resistant to antifolates [30].

Because killing of unmodified stem cells is a necessary requirement for stem cell selection, we undertook a detailed study of the mechanism for intrinsic antifolate resistance in early myeloid cells. It had previously been shown that certain tumor cells utilize exogenous nucleotide precursors to overcome the effects of antifolates in vitro [31]. We therefore investigated whether myeloid progenitors and stem cells could salvage thymidine and hypoxanthine from the serum to resist antifolate toxicity. These experiments showed that both progenitors and stem cells could efficiently utilize mucleotide salvage mechanisms to overcome TMTX resistance [32]. To sensitize the stem cell compartment to TMTX, we used drugs that inhibit nucleoside transport [33], hypothesizing that blockade of thymidine uptake would sensitize cells to TMTX by inducing a thymidineless death (fig. 2). As a transport inhibitor, we focused on nitrobenzylmercaptopurine riboside (NBMPR) based on its high affinity for blocking the facilitative *es* nucleoside transporter [35]. Our initial in vitro experiments clearly showed that NBMPR potently sensitized both murine and human myeloid progenitors to TMTX [32]. Furthermore, treatment of mice with TMTX and NBMPR-P, the latter a prodrug of NBMPR, resulted in significant depletion of normal myeloid progenitors

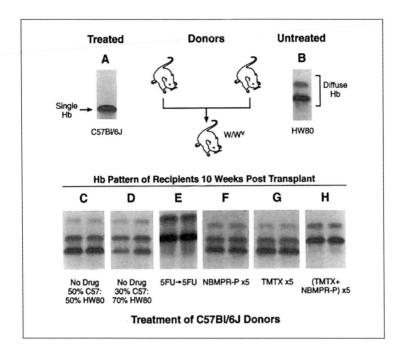

Fig. 4. Depletion of normal stem cells by treating mice with TMTX and NBMPR-P. C57Bl/6J mice were treated with a variety of drugs indicated on the bottom of panels C–H. After treatment, their bone marrow cells were mixed with equal volumes of marrow from untreated HW80 mice. These mixtures were transplanted into irradiated mice, and engraftment from each donor source was assayed at 10 weeks by hemoglobin electrophoresis. Results from 2 recipient mice are shown for each treatment group. This experiment shows that the combination of TMTX and NBMPR-P resulted in efficient killing of stem cells in treated C57Bl/6J donor mice (panel H). Reprinted from Allay et al. [32].

(fig. 3) and repopulating stem cells (fig. 4) within the bone marrow. These results demonstrated the importance of salvage mechanisms in modulating the sensitivity of early hematopoietic cells to antifolates, and defined a novel pharmacological approach for conferring selective pressure for DHFR-transduced stem cells.

Demonstration of in vivo Selection Using Bicistronic Vectors Expressing the Human L22Y DHFR Variant

For designing selectable vectors, it was first necessary to determine which DHFR cDNA would confer the greatest levels of antifolate resistance,

Table 1. Resistance to MTX and TMTX conferred on PA317 cells by cDNAs for wild-type or variant DHFR

DHFR transfected	Resistance to MTX		Resistance to TMTX	
	IC$_{50}$, nmol/l	fold	IC$_{50}$, nmol/l	fold
None	17		17.7	
G15W	18		17	
WT	53	3.1	45	2.5
F31S	62	3.6	560	32
L22R	77	4.5	1,360	76
F31W	100	5.9	102	5.7
L22F	123	7.2	82	4.6
F31G	157	9.2	1,080	60
L22Y	157	9.2	1,800	100

IC$_{50}$ drug concentration giving half as many cell colonies as in the absence of drug. Rows are arranged in order of increasing resistance to MTX. Fold resistance expressed as ratio of IC$_{50}$ for transfected cells to IC$_{50}$ for untransfected cells. Reprinted from Spencer et al. [27].

reasoning that expression could be limiting in stem cells so that adequate levels of drug resistance could be difficult to obtain. A panel of human DHFR cDNAs that contained single amino acid substitutions was screened in transfected murine fibroblasts. These studies showed that a cDNA containing a leucine to tyrosine substitution gave the highest levels of resistance to MTX and TMTX (table 1). Interestingly, the level of resistance to TMTX was an order of magnitude greater than that to MTX, presumably due to the greater degree of specificity of TMTX for DHFR inhibition.

We next constructed two bicistronic vectors containing the DHFR^{L22Y} cDNA and a transcriptionally linked reporter gene (fig. 5). The HaCID vector was based on the Harvey murine sarcoma backbone [36] and expressed both the CD24 and internal-ribosome-entry-site (IRES)-driven DHFR gene under control of the viral promoter in the 5′ long terminal repeat. Mice were transplanted with transduced bone marrow cells, and flow cytometry was used to determine the proportions of CD24-expressing peripheral blood cells present after reconstitution. After obtaining baseline marking data, mice were treated with a 5-day course to TMTX and NBMPR-P using doses previously shown to kill the majority of unmodified stem cells in the bone marrow. After recovery from myelosuppression, peripheral blood cells were again assayed for CD24 expression. Large increases in the proportion of marked red cells were seen

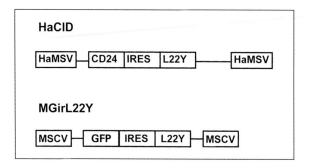

Fig. 5. Bicistronic retroviral vectors for in vivo selection experiments. Shown are the structures for the HaCID and the MGirL22Y vector. Both vectors express the L22Y DHFR variant from an IRES. The HaCID vector is based on the Harvey murine sarcoma virus (HaMSV) backbone, and contains a human CD24 cDNA as an in vivo reporter. The MGirL22Y vector is based on the MSCV and contains a cDNA for the enhanced GFP as a reporter gene.

in a number of mice following treatment (fig. 6). Several of these animals showed increases to greater than 50% marked cells despite a very low level of CD24+ erythrocytes prior to drug treatment. This was the first proof that significant enrichments for genetically modified erythrocytes could be obtained by in vivo selection, indicating a potential use of this system for gene therapy of hemoglobinopathies. Parallel increases were noted in platelets, showing that selection had occurred in multiple hematopoietic lineages (fig. 7).

To verify that selection was a specific phenomenon mediated by the DHFR resistance gene, a second group of mice were treated with 5-fluorouracil (5-FU), a drug that is not affected by DHFR expression. There was no change in the number of marked cells in these animals (fig. 7), indicating that the increases associated with TMTX and NBMPR-P were not associated with random events such as the activation of marked stem cell clones. As a further control for specificity, animals were transplanted with cells containing a vector that expressed CD24 but did not contain a DHFR gene [28]. When these mice were treated with the TMTX/NBMPR-P combination, there was no significant change in the number of marked cells further demonstrating the specificity of the selection system (fig. 7).

Improved Results with the MGirL22Y Vector

Despite the positive results obtained in these initial experiments, two limitations of the HaCID vector became apparent. The first was that some

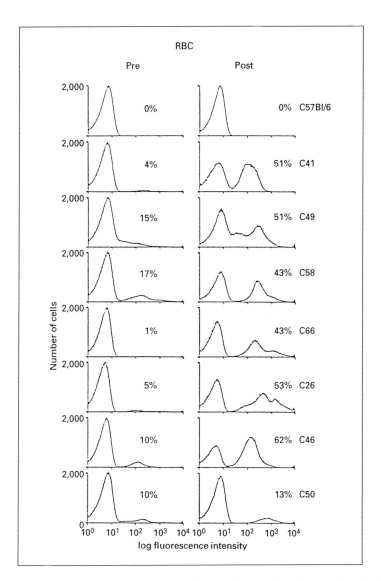

Fig. 6. Changes in CD24-expressing erythrocytes in mice transduced with the HaCID vector and treated with TMTX and NBMPR-P. Seven mice were transplanted with the HaCID vector (C26-66) and analyzed for CD24 expression in gated RBC before drug treatment (Pre). After treatment with TMTX and NBMPR-P, a second analysis was performed (Post). C57Bl/6J indicates an untransplanted control mouse. Note the large enrichments for CD24-expressing RBC that occurred in drug-treated HaCID animals.

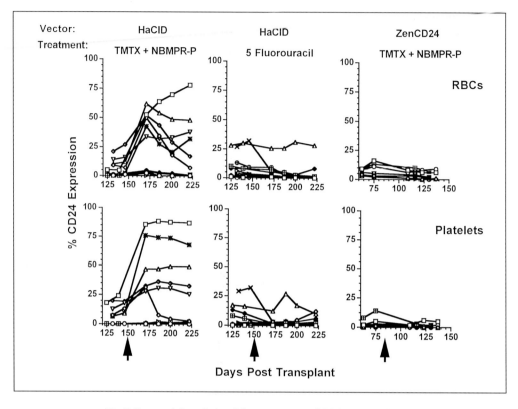

Fig. 7. Sequential analysis of the percentage of CD24-expressing peripheral blood RBC and platelets in transplanted mice treated with various drugs. Mice were transplanted with bone marrow cells transduced either with the HaCID vector (first two columns), or the ZenCD24 vector, which lacks the DHFR gene (last column). These groups were treated either with a 5-day course of TMTX and NMBPR-P, or a single dose of 5-FU as indicated. The X-axis shows the time from transplant, and the arrows indicate the day of drug treatment. Each line represents the percentage of CD24-expressing cells from a single mouse, assayed over multiple time points. The first row shows the data for RBC and the bottom row for platelets. Note that increases in marked cells were only seen in the group transplanted with HaCID-transduced cells and treated with TMTX and NMBPR-P. Reprinted from Allay et al. [40].

animals showed no increase in transduced cells with drug treatment. Because this occurred predominantly in mice with very low levels of marking prior to treatment, one possibility was that these mice were not engrafted with any transduced stem cells. A second possibility was that the Harvey murine sarcoma virus promoter was expressed at relatively low levels in primitive stem cells and that functional drug resistance was present in only a fraction of transduced stem cell clones, such as those with favorable proviral integration sites. Considering these possibilities, we next tested a second vector using the murine stem cell virus (MSCV) (fig. 5) based on prior data showing that the MSCV viral promoter was highly active in primitive stem cells [37]. We also changed the reporter gene used in the MSCV vector. Although CD24 expression from the HaCID vector was a reliable marker in red blood cells and platelets, the readout in peripheral blood leukocytes was confounded by passive transfer of the CD24 antigen to untransduced leukocytes [B.P. Sorrentino and D.A. Persons, unpubl. data]. For this reason, we next evaluated the GFP gene from jellyfish [38] and found it to be a very useful marker for hematopoietic cells. In transplanted mice and in rhesus monkeys, the GFP marker has enabled accurate tracking of marked cells in all hematopoietic lineages, including myeloid and lymphoid leukocyte subsets [29].

The resulting vector was designated MGirL22Y and expressed the GFP gene and IRES-driven $DHFR^{L22Y}$ cDNA from the retroviral promoter (fig. 5). This vector gave much higher proportions of vector-expressing cells in transplanted mice as compared with the HaCID vector [39]. In order to more closely simulate the low proportions of marked cells seen in clinical trials, MGirL22Y-transduced bone marrow cells were mixed with mock-transduced cells in a 1 to 3 ratio prior to transplantation. As expected from this dilution, between 4 and 12% of peripheral blood cells expressed the GFP marker in reconstituted mice (fig. 8). These mice were then stratified into two groups that were statistically indistinguishable regarding the level of GFP-marked cells. One group was treated with three courses of TMTX and NBMPR-P while the other control group was left untreated. In contrast to the results obtained with the HaCID vector, all mice showed significant increases in the proportion of GFP-expressing cells (fig. 8) [40]. The largest increases were seen in red cells, platelets, and in granulocytes. Lesser but statistically significant increases were seen in circulating T and B lymphocytes. These results established the ability of the DHFR selection system to enrich for vector-expressing cells from both lymphoid and myeloid lineages, and showed consistent selection among individual mice within the treatment group.

Selection with the MGirL22Y Vector Occurs at the Stem Cell Level

To establish whether selection had occurred at the level of repopulating stem cells, 5 mice were killed 6 weeks after the last drug treatment and used as marrow donors in secondary transplant recipients. Analysis of the treated animals at the time of necropsy showed enrichment for GFP-expressing pre-

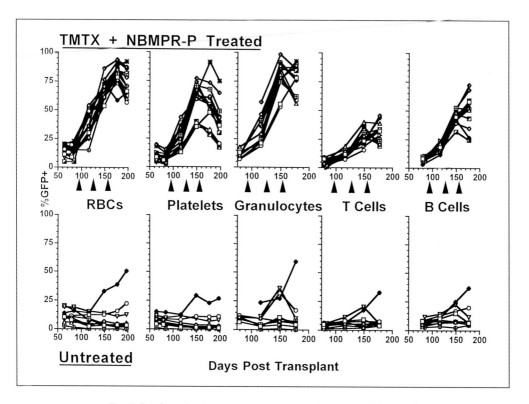

Fig. 8. In vivo selection of vector-expressing hematopoietic cells in mice transplanted with MGirL22Y-transduced bone marrow cells. Mice were transplanted with 1 part MGirL22Y-transduced cells and 3 parts mock-transduced marrow. Animals were then divided into two groups, one received 3 sequential courses of TMTX and NBMPR-P (indicated by arrows in the upper row), and the other group remained untreated as a control (lower row). Shown is the percentage of cells that expressed the GFP marker in various hematopoietic lineages (indicated for each column). For each graph, the lines represent serial data from individual mice. The increases noted for the treated group were highly significant for all lineages, but were most marked in RBC and platelets. Reprinted from Allay et al. [40].

B cells, CFU-C, and CFU-S in the bone marrow [40], consistent with selection of a very early cell.

Secondary transplant experiments were performed to definitively test whether stem cell selection had occurred (fig. 9a). Drug-treated mice were used as bone marrow donors 6 weeks after receiving the last drug treatment. As controls, mice that received no drug selection were also used as marrow donors. Bone marrow was transplanted into irradiated recipients, and marking with GFP expressing cells was analyzed for both groups over a 27-week period. If selection occurred at the stem cell level, the prediction is that the increases in GFP+ cells seen in the primary animals would be stably transmitted to the secondary recipients. In contrast, the recipients transplanted with marrow from untreated mice should have significantly fewer GFP+ hematopoietic cells. These experiments in fact showed that these predictions were true, and that selection had occurred at the stem cell level in the drug-treated donor mice (fig. 9b). In most cases, the proportion of marked cells in secondary recipients significantly exceeded 25%, the maximum number of GFP+ cells theoretically possible in the absence of selection, given the initial dilution of transduced cells with mock cells. These high levels of GFP+ marked cells were stable over the 27-week observation period, and were much higher than those seen in mice transplanted with marrow from untreated donors [40]. Therefore, these data formally demonstrate that drug treatment had enriched for stem cells transduced with the MGirL22Y vector, and provide the first proof that pluripotent hematopoietic stem cells that have been transduced with a genetic vector can be proportionally enriched with an in vivo selection system.

The Potential of the DHFR Selection System for Clinical Application

There are several critical requirements for a stem cell selection system. The drugs used must result in some degree of depletion of unmodified stem cell clones while having an acceptable toxicity profile. Secondly, the vectors used for stem cell selection must be expressed at significant levels within primitive stem cells. Lastly, the gene used as a dominant selectable marker should be stable in the context of the retroviral vector and permit the inclusion of a second therapeutic gene.

In our murine transplant model, the DHFR selection system fulfills all of these requirements. The combination of TMTX and NBMPR-P resulted in stem cell depletion as demonstrated in competitive repopulation assays. The use of a nucleoside transport inhibitor is critical for sensitization of unmodified stem cells. Intrinsic resistance of stem cells resulting from nucleotide salvage best explains the lack of selection using MTX alone in earlier experiments by Williams et al. [41]. In terms of toxicity, mice tolerated the selection regimen relatively well, with 14 of 17 mice surviving after being treated with 3 serial courses of drug selection. The main toxicity is myelosuppression and some degree of gastrointestinal toxicity. The tolerability of this regimen will have to be confirmed in large animal models, and ultimately in phase I clinical trials. Nucleotide transport inhibitors that are similar to NBMPR-P have been tested clinically as vasodilators, and appear to have few toxic side effects [42].

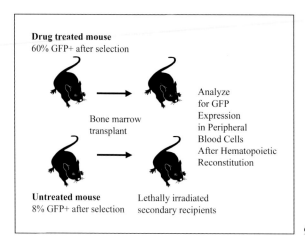

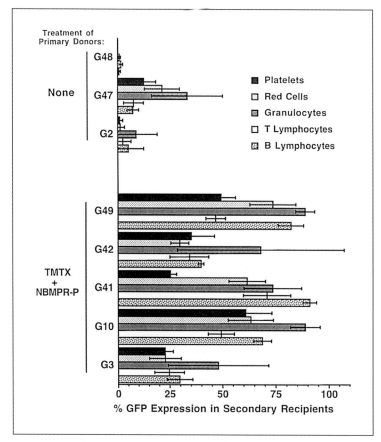

The vasodilatory effects of transport inhibitors, mediated by inhibition of adenosine uptake, can be eliminated by caffeine-induced blockade of the adenosine receptor [42].

In general, retroviral vectors are poorly expressed in primitive hematopoietic stem cells [43]. The MSCV promoter has been shown to be superior for directing expression of exogenous genes in stem cells [37], and was the basis for choosing this vector backbone for our second set of experiments. Consistent with this prediction, selection was more uniform using the MGirL22Y vector than with the HaCID vector. A second advantage is the potency of the DHFR^{L22Y} gene for conferring drug resistance. Hematopoietic cells expressing this enzyme variant were 100-fold more resistant to TMTX than control cells [27]. Therefore, even if stem cell expression were limited, the high degree of resistance conferred by this gene would compensate for low levels of gene expression. We have recently identified a DHFR variant that confers even greater levels of TMTX resistance, by virtue of a second point substitution [44]. This variant is currently being tested and may provide even greater levels of stem cell drug resistance.

Eventually, this system will have to be tested in primates to validate its utility. However, in vivo selection is not predicted to be able to overcome a total lack of engraftment with transduced stem cells. In such a case, selection could at best achieve a transient enrichment of transduced progenitors with limited self-renewal capacity, leading to only a temporary increase in marked cells. Until recently, it has not been clear whether stem cell transduction can be reproducibly achieved in nonhuman primate models [12]. Recent advances in transduction protocols are now yielding marking that is more consistent with stable transduction and engraftment of long-term repopulating cells [45, 46]. Application of these transduction protocols should allow testing of the selection hypothesis in primate models.

Our ultimate goal is to test this system in human gene therapy trials. One potential area of application is for nonmalignant diseases of the myeloid system such as chronic granulomatous disease, β-thalassemia, and sickle cell anemia. Gene transfer experiments in an animal model of chronic granuloma-

Fig. 9. Secondary transplants from drug-treated donors demonstrate in vivo selection at the stem cell level. From the experiment shown in figure 8, 3 untreated mice (G48, 47, and 2) and 5 treated mice (G49, 42, 41, 10 and 3) were used as bone marrow donors for secondary transplant experiments. Three to four lethally irradiated recipients were transplanted for each donor, and analyzed for GFP expression 27 weeks after killing the primary donor animal. *a* The average percent GFP expression is shown for various peripheral blood lineages within each transplant group. *b* Note that the predicted maximum percent GFP expression that can occur without stem cell selection is 25%. Reprinted from Allay et al. [40].

tous disease have indicated that at least 5–10% corrected cells will be necessary to positively impact the disease phenotype [7]. It is reasonable to predict that this can be achieved using stem cell selection coupled with recent advances in the transduction protocols. For the hemoglobinopathies, it may be necessary to achieve greater levels of corrected cells. Stem cell selection may also be useful for circumventing the need for myeloablative conditioning regimens. For nonmalignant diseases, it is highly desirable to avoid the considerable toxicity of these preparative treatments. The disadvantage of reducing the intensity of ablation is that much fewer transduced cells engraft [47]. It may be possible to amplify this small minority of transduced stem cells by applying serial rounds of drug selection after transplant.

Acknowledgments

This work was supported in part by the National Heart, Lung, and Blood Institute Program Project Grant No. P01 HL 53749 (BPS), The James S. McDonnell Foundation Grant No. 94-50 (BPS), The John H. Sununu Postdoctoral Fellowship (JAA), The Assisi Foundation of Memphis Grant 94-00, US Public Health Service Grant No. P01 CA 31922 (RLB), Cancer Center Support Grant No. P30 CA 21765 (BPS, RLB), and the American Lebanese Syrian Associated Charities (ALSAC).

References

1 Buckley RH, Schiff SE, Schiff RI, Roberts JL, Markert ML, Peters W, Williams LW, Ward FE: Haploidentical bone marrow stem cell transplantation in human severe combined immunodeficiency. Semin Hematol 1993;30:92.
2 Sorrentino BP, Nienhuis AW: The hematopoietic system as a target for gene therapy; in Friedmann T (ed): The Development of Gene Therapy. New York, Cold Spring Harbor Laboratory Press, in press.
3 Taylor N, Uribe L, Smith S, Jahn T, Kohn DB, Weinberg K: Correction of interleukin-2 receptor function in X-SCID lymphoblastoid cells by retrovirally mediated transfer of the gamma-c gene. Blood 1996;87:3103.
4 Candotti F, Oakes SA, Johnston JA, Notarangelo LD, O'Shea JJ, Blaese RM: In vitro correction of JAK3-deficient severe combined immunodeficiency by retroviral-mediated gene transduction. J Exp Med 1996;183:2687.
5 Candotti F, Johnston JA, Puck JM, Sugamura K, O'Shea JJ, Blaese RM: Retroviral-mediated gene correction for X-linked severe combined immunodeficiency. Blood 1996;87:3097.
6 Mardiney M, Jackson SH, Spratt SK, Li F, Holland SM, Malech HL: Enhanced host defense after gene transfer in the murine p47phox-deficient model of chronic granulomatous disease. Blood 1997; 89:2268.
7 Bjorgvinsdottir H, Ding C, Pech N, Gifford MA, Li LL, Dinauer MC: Retroviral-mediated gene transfer of gp91phox into bone marrow cells rescues defect in host defense against *Aspergillus fumigatus* in murine X-linked chronic granulomatous disease. Blood 1997;89:41.
8 Bunting KD, Sangster MY, Ihle JN, Sorrentino BP: Restoration of lymphocyte function in Janus kinase 3-deficient mice by retroviral-mediated gene transfer. Nat Med 1998;4:58.

9 Dick JE, Magli MC, Huszar D, Phillips RA, Bernstein A: Introduction of a selectable gene into primitive stem cells capable of long-term reconstitution of the hemopoietic system of W/Wv mice. Cell 1985;42:71.
10 Eglitis MA, Kantoff P, Gilboa E, Anderson WF: Gene expression in mice after high efficiency retroviral-mediated gene transfer. Science 1985;230:1395.
11 Keller G, Paige C, Gilboa E, Wagner EF: Expression of a foreign gene in myeloid and lymphoid cells derived from multipotent haematopoietic precursors. Nature (Lond) 1985;318:149.
12 Van Beusechem VW, Valerio D: Gene transfer into hematopoietic stem cells of nonhuman primates. Hum Gene Ther 1996;7:1649.
13 Dunbar CE, Cottler-Fox M, O'Shaughnessy JA, Doren S, Carter C, Berenson R, Brown S, Moen RC, Greenblatt J, Stewart FM, et al: Retrovirally marked CD34-enriched peripheral blood and bone marrow cells contribute to long-term engraftment after autologous transplantation. Blood 1995;85:3048.
14 Brenner MK, Rill DR, Holladay MS, Heslop HE, Moen RC, Buschle M, Krance RA, Santana VM, Anderson WF, Ihle JN: Gene marking to determine whether autologous marrow infusion restores long-term haemopoiesis in cancer patients. Lancet 1993;342:1134.
15 Hanania EG, Giles RE, Kavanagh J, Ellerson D, Zu Z, Wang T, Su Y, Kudelka A, Rahman Z, Holmes F, Hortobagyi G, Claxton D, Bachier C, Thall P, Cheng S, Hester J, Ostrove JM, Bird RE, Chang A, Korbling M, Seong D, Cote R, Holzmayer T, Mechetner E, Deisseroth AM, et al: Results of MDR-1 vector modification trial indicate that granulocyte/macrophage colony-forming unit cells do not contribute to posttransplant hematopoietic recovery following intensive systemic therapy. Proc Natl Acad Sci USA 1996;93:15346.
16 Hesdorffer C, Ayello J, Ward M, Kaubisch A, Vahdat L, Balmaceda C, Garrett TD, Fetell M, Reiss R, Bank A, Antman K: Phase I trial of retroviral-mediated transfer of the human MDR1 gene as marrow chemoprotection in patients undergoing high-dose chemotherapy and autologous stem-cell transplantation. J Clin Oncol 1998;16:165.
17 Orlic D, Girard LJ, Jordan CT, Anderson SM, Cline AP, Bodine DM: The level of mRNA encoding the amphotropic retrovirus receptor in mouse and human hematopoietic stem cells is low and correlates with the efficiency of retrovirus transduction. Proc Natl Acad Sci USA 1996;93: 11097.
18 Miller DG, Adam MA, Miller AD: Gene transfer by retrovirus vectors occurs only in cells that are actively replicating at the time of infection. Mol Cell Biol 1990;10:4239.
19 Kohn DB, Hershfield MS, Carbonaro D, Shigeoka A, Brooks J, Smogorzewska EM, Barsky LW, Chan R, Burotto F, Annett G, Nolta JA, Crooks GM, Kapoor N, Elder M, Ara D, Owen T, Madsen E, Snyder FF, Bastian J, Muul L, Blaese RM, Weinberg K, Parkman R: T lymphocytes with a normal ADA gene accumulate after transplantation of transduced autologous umbilical cord blood CD34+ cells in ADA-deficient SCID neonates. Nat Med 1998;4:775.
20 Sorrentino BP: Drug resistance gene therapy; in Brenner MK, Moen RC (eds): Gene Therapy in Cancer. New York, Marcel Dekker, 1996, p 189.
21 Sorrentino BP, Brandt SJ, Bodine D, Gottesman M, Pastan I, Cline A, Nienhuis AW: Selection of drug-resistant bone marrow cells in vivo after retroviral transfer of human MDR1. Science 1992; 257:99.
22 Podda S, Ward M, Himelstein A, Richardson C, de la Flor-Weiss E, Smith L, Gottesman M, Pastan I, Bank A: Transfer and expression of the human multiple drug resistance gene into live mice. Proc Natl Acad Sci USA 1992;89:9676.
23 Allay JA, Davis BM, Gerson SL: Human alkyltransferase-transduced murine myeloid progenitors are enriched in vivo by BCNU treatment of transplanted mice. Exp Hematol 1997;25: 1069.
24 Davis BM, Reese JS, Koc ON, Lee K, Schupp JE, Gerson SL: Selection for G156A O^6-methylguanine DNA methyltransferase gene-transduced hematopoietic progenitors and protection from lethality in mice treated with O^6-benzylguanine and 1,3-bis(2-chloroethyl)-1-nitrosourea. Cancer Res 1997; 57:5093.
25 Simonsen CC, Levinson AD: Isolation and expression of an altered mouse dihydrofolate reductase cDNA. Proc Natl Acad Sci USA 1983;80:2495.

26 Blakley RL, Sorrentino BP: In vitro mutations in dihydrofolate reductase that confer resistance to methotrexate: Potential for clinical application. Hum Mutat 1998;11:259.
27 Spencer HT, Sleep SE, Rehg JE, Blakley RL, Sorrentino BP: A gene transfer strategy for making bone marrow cells resistant to trimetrexate. Blood 1996;87:2579.
28 Pawliuk R, Kay R, Lansdorp P, Humphries RK: Selection of retrovirally transduced hemotopoietic cells using CD24 as a marker of gene transfer. Blood 1994;84:2868.
29 Persons DA, Allay JA, Riberdy JM, Wersto RP, Donahue RE, Sorrentino BP, Nienhuis AW: Utilization of the green fluorescent protein gene as a marker to identify and track genetically-modified hemotopoietic cells. Nat Med 1998;4:1201.
30 Blau CA, Neff T, Papayannopoulou T: The hematological effects of folate analogs: Implications for using the dihydrofolate reductase gene for in vivo selection. Hum Gene Ther 1996;7:2069.
31 Belt JA, Marina NM, Phelps DA, Crawford CR: Nucleoside transport in normal and neoplastic cells. Adv Enzyme Regul 1993;33:235.
32 Allay JA, Spencer HT, Wilkinson SL, Belt JA, Blakley RL, Sorrentino BP: Sensitization of hematopoietic stem and progenitor cells to trimetrexate using nucleoside transport inhibitors. Blood 1997;90:3546.
33 Cass CE: Nucleoside transport; in Georgopapadakou NH (ed): Drug Transport in Antimicrobial Therapy and Anticancer Therapy. New York, Marcel Dekker, 1995, p 403.
34 Allay JA, Galipeau J, Blakley RL, Sorrentino BP: Retroviral vectors containing a variant dihydrofolate reductase gene for drug protection and in vivo selection of hematopoietic cells. Stem Cells 1998;16(suppl 1):223.
35 Kolassa N, Jakobs ES, Buzzell GR, Paterson AR: Manipulation of toxicity and tissue distribution of tubercidin in mice by nitrobenzylthioinosine 5′-monophosphate. Biochem Pharmacol 1982;31:1863.
36 Pastan I, Gottesman MM, Ueda K, Lovelace E, Rutherford AV, Willingham MC: A retrovirus carrying an MDR1 cDNA confers multidrug resistance and polarized expression of P-glycoprotein in MDCK cells. Proc Natl Acad Sci USA 1988;85:4486.
37 Hawley RG, Lieu FH, Fong AZ, Hawley TS: Versatile retroviral vectors for potential use in gene therapy. Gene Ther 1994;1:136.
38 Chalfie M, Tu Y, Euskirchen G, Ward WW, Prasher DC: Green fluorescent protein as a marker for gene expression. Science 1994;263:802.
39 Persons DA, Allay JA, Allay ER, Smeyne RJ, Ashmun RA, Sorrentino BP, Nienhuis AW: Retroviral-mediated transfer of the green fluorescent protein gene into murine hematopoietic cells facilitates scoring and selection of transduced progenitors in vitro and identification of genetically modified cells in vivo. Blood 1997;90:1777.
40 Allay JA, Persons DA, Galipeau J, Riberdy JM, Ashmun RA, Blakley RL, Sorrentino BP: In vivo selection of retrovirally transduced hematopoietic stem cells. Nat Med 1998;4:1136.
41 Corey CA, DeSilva AD, Holland CA, Williams DA: Serial transplantation of methotrexate-resistant bone marrow: Protection of murine recipients from drug toxicity by progeny of transduced stem cells. Blood 1990;75:337.
42 Rongen GA, Smits P, Ver Donck K, Willemsen JJ, De Abreu RA, Van Belle H, Thien T: Hemodynamic and neurohumoral effects of various grades of selective adenosine transport inhibition in humans. Implications for its future role in cardioprotection. J Clin Invest 1995;95:658.
43 Beck-Engeser G, Stocking C, Just U, Albritton L, Dexter M, Spooncer E, Ostertag W: Retroviral vectors related to the myeloproliferative sarcoma virus allow efficient expression in hematopoietic stem and precursor cell lines, but retroviral infection is reduced in more primitive cells. Hum Gene Ther 1991;2:61.
44 Patel M, Sleep SE, Lewis WS, Spencer HT, Mareya SM, Sorrentino BP, Blakley RL: Comparison of the protection of cells from antifolates by transduced human dihydrofolate reductase mutants. Hum Gene Ther 1997;8:2069.
45 Tisdale JF, Hanazono Y, Sellers SE, Agricola BA, Metzger ME, Donahue RE, Dunbar CE: Ex vivo expansion of genetically marked rhesus peripheral blood progenitor cells results in diminished long-term repopulating ability. Blood 1998;92:1131.

46 Kiem HP, Andrews RG, Morris JA, Peterson L, Heyward S, Allen JM, Rasko JE, Potter J, Miller AD: Improved gene transfer into baboon marrow repopulating cells using recombinant human fibronectin fragment CH-296 in combination with interleukin-6, stem cell factor, FLT-3 ligand, and megakaryocyte growth and development factor. Blood 1998;92:1878.

47 Malech HL, Maples PB, Whiting-Theobald N, Linton GF, Sekhsaria S, Vowells SJ, Li F, Miller JA, DeCarlo E, Holland SM, Leitman SF, Carter CS, Butz RE, Read EJ, Fleisher TA, Schneiderman RD, Van Epps DE, Spratt SK, Maack CA, Rokovich JA, Cohen LK, Gallin JI: Prolonged production of NADPH oxidase-corrected granulocytes after gene therapy of chronic granulomatous disease. Proc Natl Acad Sci USA 1997;94:12133.

Brian P. Sorrentino, MD, Assistant Faculty Member, Department of Biochemistry,
St. Jude Children's Research Hospital, Memphis, TN 38105 (USA)
Tel. +1 901 495 2727, Fax +1 901 495 2176, E-Mail brian.sorrentino@stjude.org

In vivo Selection of Genetically Modified Bone Marrow Cells

C. Anthony Blau

Division of Hematology, University of Washington, Seattle, Wash., USA

Inefficient gene transfer into the human hematopoetic stem cell presents what is arguably the single most difficult obstacle confronting stem cell gene therapy [1]. Gene marking trials indicate that fewer than 1% of stem cells incorporate the marker gene [2, 3]. Therapeutic applications for stem cell gene transfer require methods for markedly increasing the frequency of genetically corrected stem cells. A strategy for increasing the proportion of modified stem cells is to specifically direct their preferential expansion through selection [4]. Bicistronic vectors allow a gene encoding a selectable product to be linked to a gene encoding a therapeutic protein; following selection, cells bearing the therapeutic gene emerge. Ex vivo selection of genetically modified primary hematopoietic cells has been achieved using drug resistance genes [5], or genes encoding surface membrane proteins [6]. However, two problems impose major limitations on the utility of ex vivo selection. First, a very large number of nontransduced stem cells residing in the host will compete with the relatively small number of transduced stem cells to be infused, unless myeloablation is applied. Second, extensive manipulations of cells associated with ex vivo selection may be accompanied by the loss of engraftment potential [7]. The ability to extend selection to the in vivo setting may allow these problems to be overcome. Furthermore, if a clinically tolerable strategy for in vivo selection were devised, selection could be applied repeatedly.

Current approaches for in vivo selection involve the transfer of a drug resistance gene into a minor population of hematopoietic cells. Selective pressure is applied through the in vivo administration of the appropriate cytotoxic drug. Success requires that the cytotoxic drug exert a proportionally greater toxic effect on the population of unmodified marrow cells relative to their transduced counterparts. Additionally, if the effects of selection are to persist,

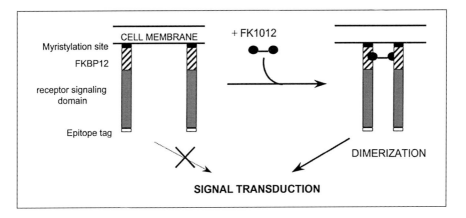

Fig. 1. Schematic diagram of FK1012-mediated signal transduction. The myristylation domain targets the fusion protein to the inner surface of the cell membrane. The FKBP12 domain provides a binding site for the membrane permeable drug FK506 which is dimerized to form the compound FK1012 [25]. In the absence of FK1012, the fusion protein is unable to dimerize and is therefore unable to initiate signal transduction. Addition of FK1012 results in dimerization of the fusion protein thereby triggering signal transduction.

it is anticipated that selection must occur minimally at the level of early hematopoietic progenitors, and ideally at the level of stem cells. The genes most frequently used for this application are dihydrofolate reductase (DHFR), which confers resistance to folate analogues [8], and multiple drug resistance gene 1 (MDR1), which confers resistance to naturally occurring drugs including taxol, navelbine and vinblastine [9]. More recently developed dominant selectable markers include O^6-alkylguanine DNA alkyltransferase [10], which confers resistance to the nitrosoureas, and cytidine deaminase [11, 12], which confers resistance to cytosine arabinoside and gemcitabine.

Recent studies in our laboratory have uncovered a major problem in using the DHFR and MDR1 genes for in vivo selection: early hematopoietic cells tolerate very high dosages of the chemotherapy drugs to which these genes confer resistance [13, 14]. In other words, chemotherapy provides little or no selective advantage to progenitors bearing the drug resistance gene, since these early hematopoietic cells are normally highly drug resistant [13–16]. The resistance of normal hematopoietic stem cells and progenitors to killing by these cytotoxic agents provides a likely explanation for the limited success achieved in using DHFR and MDR1 for in vivo selection [17–20].

Several approaches for overcoming progenitor resistance to cytotoxic drugs have been pursued. The resistance of progenitors and stem cells to many cytotoxic drugs is attributable to their infrequent entry into cell cycle [21].

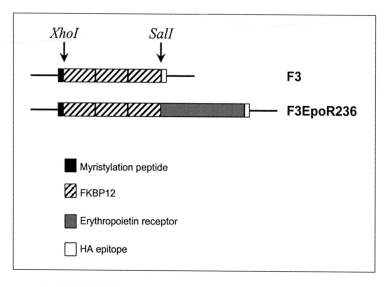

Fig. 2. FKBP12/EpoR constructs. The construct F3 encodes a fusion protein containing a myristylation domain to direct localization to the inner surface of the cell membrane, 3 copies of FKBP12 which binds the drug FK506, and an HA epitope tag, to permit detection of the protein using a commercially available antibody. F3EpoR236 contains the full length 236 amino acid EpoR cytoplasmic domain inserted into the *Sal*I site of F3.

Progenitors can be driven into cell cycle through the use of cytokines, thereby increasing their susceptibility to many chemotherapeutic agents. Experiments by Molineux et al. [22] and Harrison et al. [23] have shown that the preceding administration of c-*kit* ligand (stem cell factor; SCF) markedly accentuates progenitor killing by 5-fluorouracil (5-FU). In analogous experiments, we have shown that SCF also sensitizes progenitors to killing by many of the drugs to which MDR1 confers resistance, including navelbine, vinblastine and taxol [14]. Also, flt-3 ligand (FL) shares the ability of SCF to sensitize progenitors to 5-FU [manuscript submitted]. In contrast, the resistance of progenitor and stem cells to methotrexate and trimetrexate cannot be overcome by stimulating entry into cell cycle [13]; rather cellular resistance to these drugs can be overcome by blocking uptake of nucleotide precursors [24]. Finally, in contrast to most chemotherapeutic agents, a handful of drugs, including the nitrosoureas, are capable of exerting toxicity at the level of stem cells. Accordingly, forced expression of the enzyme O^6-alkylguanine DNA alkyltransferase in hematopoietic progenitor cells allows for their in vivo selection using 1,3-bis-(2-chloroethyl) nitrosourea [10].

An alternative approach for accomplishing in vivo selection would be to confer a direct proliferative advantage upon the genetically modified cell

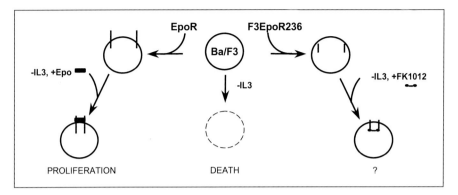

Fig. 3. Experimental design used to demonstrate that FK1012 is able to activate proliferative signaling. Ba/F3 cells are a mouse pro-B cell line. In the absence of IL-3, cells die over a period of 2–3 days. In contrast, cells transfected with a construct encoding the EpoR are able to proliferate, despite IL-3 withdrawal, in the presence of erythropoietin. Ba/F3 cells expressing the F3EpoR236 fusion protein are able to proliferate in an FK1012-dependent manner [30].

population. The practical application of this strategy would require that the proliferative stimulus: (1) by restricted to the genetically modified population, and (2) be completely reversible. We have exploited a new method that appears to meet these requirements. This approach is based on the principle that many cytokine receptors are composed of single chains that are activated upon ligand-induced homodimerization. Recently developed technology allows intracellular protein dimerization to be reversibly activated in response to a lipid-soluble dimeric form of the drug FK506, called FK1012 [25]. FK1012 is used as a pharmacological mediator of dimerization to bring together two FK506 binding domains, taken from the endogenous protein FKBP12. Thus as shown in figure 1, fusion proteins consisting of a cytokine-receptor-signaling domain linked to an FKBP12 domain may be dimerized and thereby activated using FK1012 [25]. This approach has been used to activate apoptosis through the Fas-signaling pathway [26] and a related approach has been used to inducibly activate synthesis of a reporter gene in vivo [27].

To demonstrate the feasibility of this system for reversibly inducing cell proliferation, a fusion protein containing three copies of the FKBP12 motif linked to the signaling domain of the erythropoietin receptor (F3EpoR236) was tested (fig. 2). The presence of a myristylation domain targets the fusion protein to the inner surface of the cell membrane, and an epitope tag permits the fusion protein to be readily detected by Western analysis. Testing was performed using IL-3-dependent murine pro-B cell line, Ba/F3 (fig. 3) [28].

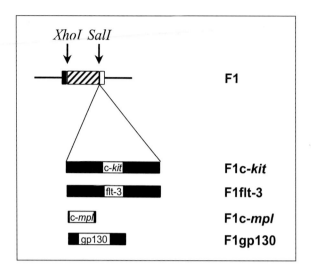

Fig. 4. Candidate molecules for stem cell expansion. The construct F1 is the same as F3 except that it contains only a single FKBP12 motif. F1c-*kit* contains the full length 432 amino acid c-*kit* cytoplasmic domain inserted into the *Sal*I site of F1, F1flt-3 contains the full length 438 amino acid cytoplasmic domain of the flt-3 receptor, F1c-*mpl* contains the full length 121 amino acid cytoplasmic portion of *mpl*, and F1gp130 contains the full length 277 amino acid cytoplasmic portion of gp130. For key to color codes, please see figure 2.

Ectopic expression of a functional erythropoietin receptor (EpoR) in Ba/F3 cells allows erythropoietin to substitute for the requirement for IL-3 [29]. In Ba/F3 cell clones expressing the F3EpoR236 fusion protein, cell growth could be sustained indefinitely using FK1012, despite the absence of IL-3 [30]. Specific inhibition of the proliferative effect of FK1012 was observed in the presence of the competing monomer, FK506 [30] (data not shown).

Toward our goal of developing a method that can be applied to the expansion of genetically modified hematopoietic stem cells, we have tested signaling molecules that are candidates for having the capacity to cause stem cells to divide: c-*kit*, flt-3, *mpl* and gp130 (fig. 4). Recent studies have shown that fusion proteins consisting of FKBP12 linked to the intracellular portions of each of these receptors allow Ba/F3 cells to proliferate in response to FK1012 [31, 32].

Considerable progress has been made in evaluating fusion proteins containing the F1mpl construct (fig. 4). Mpl, the receptor for thrombopoietin [33–35], was selected because of the demonstrated capacity of thrombopoietin for stimulating proliferation in multilineage hematopoietic progenitors in vivo [36] and because of the small size of the mpl-signaling domain, which facilitates incorporation into retroviral vectors. Ba/F3 cells expressing F1mpl construct exhibited

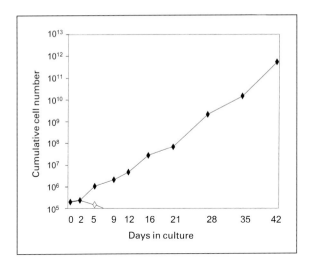

Fig. 5. FK1012 stimulates expansion of genetically modified bone marrow cells. Following retroviral transduction, marrow cells were cultured in IMDM containing 10% FCS, in the presence or absence of FK1012 (100 nM). Cells were counted at various times during culture. Open diamonds: no FK1012, closed diamonds: FK1012 alone. Note that the X axis is on a logarithmic scale. By 42 days in culture, a greater than 2.5-million-fold expansion in cell numbers had occurred.

dose-dependent proliferative responses in the presence of FK1012 (data not shown).

In order to test whether FK1012-mediated activation of mpl is capable of inducing proliferation in genetically modified bone marrow cells, the F1mpl construct was inserted into the retroviral vector MSCV-neoEB [37]. The ecotropic retrovirus MSCVF1mpl was capable of conferring G418 resistance and FK1012 responsiveness in transduced Ba/F3 cells (data not shown), and in primary murine bone marrow cells (fig. 5). Following transduction, marrow cells cultured in the absence of added growth factors died within 7–14 days. In contrast, FK1012 dramatically stimulated cell proliferation. By 42 days in culture, marrow cells had expanded 2.5-million-fold in one experiment (fig. 5) and nearly 500,000-fold in a second experiment (data not shown).

Progenitor assays (as colony-forming unit cells; CFU-C) during culture demonstrated that FK1012 stimulated the proliferation of genetically modified progenitor cells (fig. 6). As the MSCVF1mpl vector incorporates a neo gene [37], the frequency of genetically modified progenitors is reflected in the frequency of G418-resistant CFU-C. In the absence of FK1012, CFU-C expansion failed to occur, and there was no preferential survival advantage in favor of G418-resistant CFU-C. By day 28, CFU-C had fallen to undetectable levels

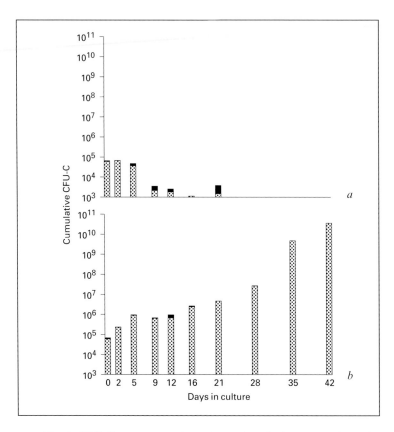

Fig. 6. CFU-C assays at various time points during suspension culture. Cells were harvested on the days indicated and cultured in semisolid media containing IL-3, either in the presence or absence of G418. The concentration of G418, 800 µg/ml, was sufficient to prevent the growth of nontransduced cells (data not shown). Dark bars: CFU-C numbers in the absence of G418; Light bars: CFU-C numbers in the presence of G418. *a* Results from suspension cultures without FK1012. *b* Results from suspension cultures in the presence of FK1012, 100 n*M*. Note that FK1012-mediated CFU-C expansion markedly favors the genetically modified cell population.

(fig. 6a). In contrast, in the presence of FK1012 an almost 600,000-fold expansion in CFU-C was observed by 42 days of culture (fig. 6b). The vast majority of CFU-C were G418 resistant, demonstrating a preferential proliferative advantage in favor of the genetically modified progenitor cell population.

In order to demonstrate that the proliferative effect of FK1012 is reversible, cells cultured for 28 days in the presence of FK1012 were tested for persistence of cell growth following FK1012 withdrawal. As shown in table 1, withdrawal

Table 1. Reversibility of FK1012-dependent cell proliferation

Growth conditions	Day in culture		
	0	7	14
+FK1012	5	36	1,283
−FK1012	5	0.59	0.09

After 28 days of culture in FK1012 (100 nM), transduced marrow cells were washed extensively and cultured in suspension in the presence (+) or absence (−) of FK1012. Numbers indicate cells per well × 10^5.

of FK1012 was followed promptly by cell death, while the readdition of FK1012 was associated with persistent cell growth.

In summary, while in vivo selection presents an intellectually appealing method for increasing the frequency of genetically modified progenitors and stem cells, the actual value of this approach remains to be demonstrated. Recent progress has arisen from the development of new selection systems, and from the increasing recognition that to be effective, selection must be imposed at the level of the stem cell.

References

1   Brenner MK: Gene transfer into hematopoietic cells. N Engl J Med 1996;335:337–339.
2   Brenner MK, Rill DR, Holladay MS, Heslop HE, Moen RC, Buschle M, Krance RA, Santana VM, Anderson WF, Ihle JN: Gene marking to determine whether autologous marrow infusion restores long-term haemopoiesis in cancer patients. Lancet 1993;342:1134–1137.
3   Dunbar CE, Cottler-Fox M, O'Shaughnessy JA, Doren S, Carter C, Berenson R, Brown S, Moen RC, Greenblatt J, Stewart FM, Leitman SF, Wilson WH, Cowan K, Young NS, Nienhuis AW: Retrovirally marked CD34-enriched peripheral blood and bone marrow cells contribute to long-term engraftment after autologous transplantation. Blood 1995;85:3048–3057.
4   Sorrentino BP, Brandt SJ, Bodine D, Gottesman M, Pastan I, Cline A, Nienhuis AW: Selection of drug-resistant bone marrow cells in vivo after retroviral transfer of human MDR1. Science 1992;257:99–103.
5   Aran JM, Gottesman MM, Pastan I: Drug-selected coexpression of human glucocerebrosidase and P-glycoprotein using a bicistronic vector. Proc Natl Acad Sci USA 1994;91:3176–3180.
6   Migita M, Medin JA, Pawliuk R, Jacobson S, Nagle JW, Anderson S, Amiri M, Humphries RK, Karlsson S: Selection of transduced CD34+ progenitors and enzymatic correction of cells from Gaucher patients, with bicistronic vectors. Proc Natl Acad Sci USA 1995;92:12075–12079.
7   Peters SO, Kittler EL, Ramshaw HS, Quesenberry PJ: Ex vivo expansion of murine marrow cells with interleukin-3 (IL-3), IL-6, IL-11, and stem cell factor leads to impaired engraftment in irradiated hosts. Blood 1996;87:30–37.
8   Simonsen CC, Levinson AD: Isolation and expression of an altered mouse dihydrofolate reductase cDNA. Proc Natl Acad Sci USA 1983;80:2495–2499.

9　Gottesman MM, Pastan I: Biochemistry of multidrug resistance mediated by the multidrug transporter. Annu Rev Biochem 1993;62:385–427.
10　Allay JA, Davis BM, Gerson SL: Human alkyltransferase-transduced murine myeloid progenitors are enriched in vivo by BCNU treatment of transplanted mice. Exp Hematol 1997;25:1069–1076.
11　Neff T, Blau CA: Forced expression of cytidine deaminase confers resistance to cytosine arabinoside and gemcitabine. Exp Hematol 1996;24:1340–1346.
12　Momparler RL, Laliberté J, Eliopoulos N, Beauséjour C, Cournoyer D: Transfection of murine fibroblast cells with human cytidine deaminase cDNA confers resistance to cytosine arabinoside. Anticancer Drugs 1996;7:266–274.
13　Blau CA, Neff T, Papayannopoulou Th: The hematological effects of folate analogs: Implications for using the dihydrofolate reductase gene for in vivo selection. Hum Gene Ther 1996;7:2069–2078.
14　Blau CA, Neff T, Papayannopoulou Th: Cytokine prestimulation as a gene therapy strategy: Implications for using the MDR1 gene as a dominant selectable marker. Blood 1997;89:146–154.
15　Stromhaug A, Warren DJ, Slordal L: Effect of methotrexate on murine bone marrow cells in vitro: Evidence of a reversible antiproliferative action. Exp Hematol 1995;23:439–443.
16　Stromhaug A, Slordal L, Warren DJ: Differential effects of the antifolates methotrexate, aminopterin and trimetrexate on murine haemopoietic progenitor cells. Br J Hematol 1996;92:514–520.
17　Stead RB, Kwok WW, Storb R, Miller AD: Canine model for gene therapy: Inefficient gene expression in dogs reconstituted with autologous marrow infected with retroviral vectors. Blood 1988;71:742–747.
18　Corey CA, DeSilva AD, Holland CA, Williams DA: Serial transplantation of methotrexate-resistant bone marrow: Protection of murine recipients from drug toxicity by progeny of transduced stem cells. Blood 1990;75:337–343.
19　Podda S, Ward M, Himelstein A, Richardson C, de la Flor-Weiss E, Smith L, Gottesman M, Pastan I, Bank A: Transfer and expression of the human multiple drug resistance gene into live mice. Proc Natl Acad Sci USA 1992;89:9676–9680.
20　Sorrentino BP, Brandt SJ, Bodine D, Gottesman M, Pastan I, Cline A, Nienhuis AW: Selection of drug-resistant bone marrow cells in vivo after retroviral transfer of human MDR1. Science 1992;257:99–103.
21　Fleming WH, Alpern EJ, Uchida N, Ikuta K, Weissman IL: Functional heterogeneity is associated with the cell cycle status of murine hematopoietic stem cells. J Cell Biol 1993;122:897–902.
22　Molineux G, Migdalska A, Haley J, Evans GS, Dexter TM: Total marrow failure induced by pegylated stem-cell factor administration before 5-fluorouracil. 1994;83:3491–3499.
23　Harrison DE, Zsebo DM, Astle CM: Splenic primitive hematopoietic stem cell (PHSC) activity is enhanced by steel factor because of PHSC proliferation. Blood 1994;83:3146–3151.
24　Allay JA, Spencer HT, Wilkinson SL, Belt JA, Blakley RL, Sorrentino BP: Sensitization of hematopoietic stem and progenitor cells to trimetrexate using nucleoside transport inhibitors. Blood 1997;90:3546–3554.
25　Spencer DM, Wandless TJ, Schreiber SL, Crabtree GR: Controlling signal transduction with synthetic ligands. Science 1993;262:1019–1034.
26　Spencer DM, Belshaw PJ, Chen L, Ho SN, Randazzo F, Crabtree GR, Schreiber S: Functional analysis of Fas signaling in vivo using synthetic inducers of dimerization. Curr Biol 1996;6:839–847.
27　Rivera V, Clackson T, Natesan S, Pollock R, Amara JF, Keenan T, Magari SR, Phillips T, Courage NL, Cerasoli F, Holt DA, Gilman M: A humanized system for pharmacologic control of gene expression. Nat Med 1996;2:1028–1032.
28　Palacios R, Steinmetz M: IL3-dependent mouse clones that express B-220 surface antigen, contain Ig genes in germ-line configuration, and generate B lymphocytes in vivo. Cell 1985;41:727–734.
29　Li JP, D'Andrea AD, Lodish HF, Baltimore D: Activation of cell growth by binding of Friend spleen focus-forming virus gp55 glycoprotein to the erythropoietin receptor. Nature 1990;343:762–764.
30　Blau CA, Peterson KR, Drachman JG, Spencer DM: A proliferation switch for genetically modified cells. Proc Natl Acad Sci USA 1997;94:3076–3081.
31　Jin L, Asano H, Blau CA: Stimulating cell proliferation through the pharmacologic activation of c-*kit*. Blood 1998;91:890–897.

32 Jin L, Siritanaraktul N, Emery DW, Richard RE, Kaushansky K, Papayannopoulou Th, Blau CA: Targeted expansion of genetically modified bone marrow cells. Proc Natl Acad Sci USA, in press.
33 Lok S, Kaushansky K, Holly RD, Kuijper JL, Lofton-Day CD, Oort PJ, Grant FJ, Hepel MD, Burkhead SK, Kramer JM, Bell LA, Sprecher CA, Blumberg J, Johnson R, Pronkard D, Ching AFT, Mathewes SL, Balley MC, Forstrom JW, Buddle MM, Osborn SG, Evans SJ, Sheppard PO, Presnell SR, O'Hara PJ, Hagen FS, Roth GJ, Foster DC: Cloning and expression of murine thrombopoietin cDNA and stimulation of platelet production in vivo. Nature 1994;369:565–568.
34 Kaushansky K, Lok S, Holly RD, Broudy VC, Lin N, Bailey MC, Forstrom JW, Buddle MM, Oort PJ, Hage RS, Roth GJ, Papayannopoulou T, Foster DC: Promotion of megakaryocyte progenitor expansion and differentiation by the c-Mpl ligand thrombopoietin. Nature 1994;369:568–571.
35 Wendling F, Maraskovsky E, Debili N, Florindo C, Teepe M, Titeux M, Methia N, Breton-Gorius J, Cosman D, Vainchenker W: c-mpl ligand is a humoral regulator of megakaryocytopoiesis. Nature 1994;369:571–574.
36 Kaushansky K, Lin N, Grossmann A, Humes J, Sprugel KH, Broudy VC: Thrombopoietin expands erythroid, granulocyte-macrophage, and megakaryocytic progenitor cells in normal and myelosuppressed mice. Exp Hematol 1996;24:265–269.
37 Hawley RG, Lieu FHL, Fong AZC, Hawley TS: Versatile retroviral vectors for potential use in gene therapy. Gene Ther 1994;1:136–138.

Dr. Anthony Blau, Division of Hematology, Mail Stop 357720, Health Sciences Bldg., University of Washington, 1959 Pacific Ave. S.E., Seattle, WA 98195 (USA)
Tel. 001 206 685 6873, Fax 001 206 543 3560, E-Mail tblau@u.washington.edu

Gene Transfer in the Nonmyeloablated Host

Peter J. Quesenberry, Pamela S. Becker

University of Massachusetts Medical Center,
University of Massachusetts Cancer Center, Worcester, Mass., USA

In order for lymphohematopoietic stem cells to engraft in vivo, dogma holds that some type of myeloablative treatment, usually irradiation, is necessary to 'open' space. This dogma has prevailed despite work dating to the 60s and 70s showing some levels of engraftment into nonmyeloablated mice [1–3]. Brecher et al. [4] demonstrated relatively high levels of engraftment when 40 million male BALB/c cells were infused daily for 5 days into BALB/c female mice. In a similar fashion engraftment into normal nonmyeloablated mice, albeit at different levels, was demonstrated by Saxe et al. [5]. The data of Brecher et al. [4] were confirmed and extended by Stewart et al. [6] who showed high levels of multilineage engraftment of male BALB/c cells into nonmyeloablated female BALB/c hosts over 2 years after marrow infusion. Further work has demonstrated that the cell which engrafts into nonmyeloablated mice is noncycling, or dormant, as determined by hydroxyurea suicide [7]. In addition, it was established that the schedule of cell injections was not critical; the same level of engraftment into nonmyeloablated hosts was obtained when 200 million cells were given over 5 days in divided (40 million per injection) tail vein injections or given in one injection, although at higher cell concentrations, there did tend to be cell clumping, which could be prevented by including low levels of heparin in the cell suspension [8]. Additional studies with this model have indicated a roughly linear increase in engraftment with increasing numbers of cells given in 5 divided doses over 5 days [9]. In a similar fashion, the level of engraftment sequentially increased with the same number of cells given for 1, 2, 3, 4 or 5 days. In these studies, engraftment was monitored at 20–25 weeks after engraftment. The engraftment levels seen in nonmyeloablated mice were analyzed using mathematical models and several basic assumptions: (1) that the engrafting cells replaced or added to host cells;

Table 1. Comparison of highest individual engraftments to theoretical levels

	Observed engraftment BM %	Maximal theoretical engraftment (BM, SP, THY)[1] %	Level above 3-organ theoretical	Maximal theoretical engraftment (BM only)[2] %	Level above 1-organ theoretical
1	79	29	2.7×	51	1.5×
2	75	24	2.6×	44	1.7×
3	72	29	2.6×	51	1.4×
4	71	16	4.4×	29	2.4×
5	56	29	1.9×	51	1.1×
6	54	9	6.0×	16	3.4×
7	47	29	1.6×	51	–
8	46	9	5.1×	16	2.9×
9	44	29	1.5×	51	–
10	40	29	1.4×	51	–

Modification reproduced from Rao et al. [9].

[1] Theoretical model of distribution of infused cells between bone marrow (BM), spleen (SP) and thymus (THY).

[2] Theoretical model of distribution of infused cells homing solely to bone marrow (BM).

(2) that all cells engrafted; (3) that the end cell progeny correlated numerically with the stem cells, and (4) that total host marrow cells were approximately 300 million in number. Utilizing the replacement model, with the above assumptions, and regardless of the cell level infused and number of infusions, the rate of engraftment in marrow approached or exceeded the highest rates of engraftment estimated by the theoretical model (table 1).

These data indicate that engraftment is not dependent upon myeloablation 'opening space', but rather that both short- and long-term donor to host ratios in the engrafted model are determined by the ratio of donor to host stem cells.

Further work in that nonmyeloablated model has shown that homing to marrow occurs over a window of 19 h and is probably complete at a much earlier time point. Furthermore, high levels of engraftment in nonmyeloablated mice can be seen with highly purified lineage-negative, rhodamine low and Hoechst low (Lin$^-$rho^{lo} ho^{lo}) stem cells. An intriguing aspect of these studies is that the final level of engraftment seen with purified murine stem cells is much lower than that expected considering the starting number of marrow cells from which these stem cells were purified. This was the case whether

Table 2. Engraftment of lineage negative rho^{lo} ho^{lo} cells in nonablated hosts at 6 weeks or 6 months after engraftment

Donor Lin-rho^{lo} ho^{lo} cells, n	Expected male, %	Observed male, %	Expected, %
6 weeks			
2,600	22	1.4	6.4
5,500	46	1.2	3.3
10,000	83	0.7	0.8
10,000	83	2.2	2.6
6 months			
3,000	25	0.9	3.6

immediate homing (19 h) or longer engraftment was evaluated (table 2). This apparent discrepancy could be explained by a loss of stem cells or stem cell potential with the separation procedure. Alternatively, this could indicate the existence of a facilitator cell or cells such as has been shown in allogeneic murine transplant models. Kaufman et al. [10] demonstrated a CD8$^+$TCR$^-$ cell which facilitates allogeneic engraftment.

Treatment with 5-fluorouracil (5-FU) has been shown to induce cell cycle activation of primitive hematopoietic stem cells and to enrich for high proliferative potential colony-forming cells (on a percent basis). Accordingly, we evaluated the capacity of marrow cells harvested after 5-FU to engraft in nonmyeloablated hosts [7]. To our surprise, at 1 and 6 days after 5-FU, long-term engraftment turned out to be quite defective. This engraftment defect recovered over time. Furthermore, the cells which did engraft in this setting were noncycling, as measured by in vivo hydroxyurea suicide. These data suggested that induction of proliferation was associated with a long-term engraftment defect.

Further support for the association between proliferation and engraftment was found in studies of cytokine effects in vitro on marrow progenitor/stem cells. When BALB/c murine marrow was exposed for 48 h to IL-3, IL-6, IL-11 and steel factor in liquid culture, total cell and progenitor numbers, including high proliferative potential colony-forming cells were increased; the latter were induced from dormancy into cell cycle as determined by tritiated thymidine suicide. However, long-term engraftment (>3 weeks) was impaired [11, 12]. Work on highly purified murine BALB/c lineage negative rhodamine low and Hoescht low stem cells cultured in the same cytokine cocktail demonstrated that progress from dormancy to S phase was approximately 16–20 h and to first mitosis 36–40 h. This prolonged 1st cell cycle transit was then fol-

Table 3. Factors adversely affecting engraftment and gene transfer

1	Stem cell purification
2	5-FU pretreatment of donor
3	Cytokine exposure and cell cycle induction of donor cells

lowed by sequential 12-hour cycles, indicating extremely short G_1 phases [13]. Further work has suggested that the engraftment capacity of murine hematopoietic stem cells is impaired at late S/early G_2 stages of cell cycle and that it recovers in G_1. Thus cell purification, 5-FU pretreatment and cytokine treatment all appear to impair the total engraftment efficiency of stem cells, although the final outcome appears to be related to the phase of cell cycle of the stem cell.

We attempted to establish MDR1 gene transfer into nonablated BALB/c mice [14]. Here we utilized the same cytokine cocktail plus retroviral feeder layers to induce cycling and retroviral gene integration as per the previous studies of Sorrentino et al. [15]. We also used repetitive taxol to attempt to select gene transfected stem cells. We analyzed engraftment at 14 months after 3 taxol treatments. We found reasonable levels of engraftment of cells into nonmyeloablated female hosts. We also detected presence of the MDR1 gene using sensitive RT-PCR approaches. However, we estimated that the vast majority of engrafted cells did not contain or express the gene.

These data could be explained by promotor shut off, but an alternative suggestion is that most cells which engrafted did not carry the gene, possibly because the transduced, cycling cells failed to engraft. Either possibility is supported by our observations that the marrow cells transduced in vitro showed a high percentage of transduced HPP-CFC in in vitro clonal culture.

In toto, these data indicate that while stem cell engraftment into nonmyeloablative hosts may be quantitative, cycling long-term stem cells engraft poorly. Our disappointing results with gene transfer may have been due to poor engraftment of gene carrying cycling stem cells. Factors which may militate against good long-term gene transfer into nonmyeloablated (myeloablated) hosts are presented in table 3.

Gene transfer into nonmyeloablated canine hosts has apparently been accomplished from canine Dexter cultures [16], although this remains to be confirmed. The ideal setting for such gene transfer approaches are in nonmalignant marrow diseases.

A critical feature for any stem cell gene therapy success is the final donor to host ratio of differentiated marrow and blood cells. Since, as noted above, the ratio of host to donor stem cells determined the final differentiated cell

Table 4. Stem cell strategy for enhancing retroviral-based gene therapy in nonmyeloablated recipients

---

1   Transduce cells and then engraft them when in G_1
2   Either don't purify stem cells or add back putative facilitator cells
3   Utilize minimal myeloablation to enhance chimerism – low-dose irradiation

---

phenotype, a relatively large number of marrow cells is necessary to obtain a significant percentage of differentiated donor cells in the nonmyeloablated host. This would be a limiting factor in gene therapy approaches without myeloablation.

An approach which would nontoxically reduce host stem cells without significant myeloablation should markedly increase the donor to host ratio. This approach of minimal myelotoxicity with significant stem cell toxicity does in fact give high engraftment rates without significant myelotoxicity [17]. One hundred centigrays whole body irradiation produce moderate and transient decreases in the WBC and platelet counts with virtually no effect on marrow cellularity, but reduces long-term engraftable stem cells to less than 20% of control values. Mice given 100 cGy minimal myeloablation and 40 million marrow cells intravenously show very high rates of chimerism that sometimes approaches 100% at 2, 5 and 8 months after marrow infusion. Importantly, high rates of chimerism are seen with 10 million cells, levels which by extrapolation to human marrow, should be easily obtained in a routine clinical setting.

Altogether the above suggests a clear 'stem cell' strategy for improving gene transfer. This is outlined in table 4.

Marrow cell gene transfer without myeloablation has in fact been accomplished in at least one patient with Gaucher's disease. Heterozygote levels of glucocerebrosidase (GC) have been obtained out to over 12 months after transplantation of transfected autologous marrow [18]. A G-CSF-mobilized, CD34+-enriched population of progenitor cells was transduced with a retroviral vector encoding GC, MFG-GC, by centrifugation and transplanted to autologous recipients without prior myeloablation. The cell dose was about 2×10^6 CD34+ cells/kg. The GC transgene was present in all 3 patients studied at several months after transplantation, and there was an increase in the GC enzyme activity in these individuals as well. Moreover, in one of the transplanted patients, the enzyme levels rose to those found in heterozygous carriers, who are clinicallly asymptomatic, and the effect persisted beyond 12 months after the fourth and final transplantation. This patient has undergone a tenfold decrease in exogenous enzyme replacement without any clinical changes.

In order to examine the numbers of primitive progenitors and cell cycle status of the transduced patient cells, the MFG-GC-transduced CD34+ cells were examined for HPP-CFC number and by tritiated thymidine suicide assay [19]. The transduction procedure did not alter the number of HPP-CFC. The remarkable finding was that the centrifugation transduction did cause a dramatic change in the percent thymidine suicide of HPP-CFC colonies: there was $56 \pm 16\%$ thymidine suicide for the nontransduced cells, but only $8 \pm 10\%$ thymidine suicide for the transduced cells derived from the first day of pheresed peripheral stem cells. Thus, the retrovirally infected cells entered a nonproliferative phase which may have enhanced their ability to engraft in the nonmyeloablated hosts.

Overall, these data suggest that effective gene transfer in patients with various nonmalignant diseases might be achieved by nonmyelotoxic but stem cell toxic (i.e. low level irradiation) host treatment combined with approaches to quiesce transduced stem cells, thus enhancing their long-term engraftability.

References

1 Micklem HS, Clark CM, Evans EP, Ford CE: Fate of chromosome-marked mouse bone marrow cells transfused into normal syngeneic recipients. Transplantation 1968;6:299–302.
2 Takada A, Takada Y, Ambrus JL: Proliferation of donor spleen and bone-marrow cells in the spleens and bone marrow of unirradiated and irradiated adult mice. Proc Soc Exp Biol Med 1971; 136:222–226.
3 Takada Y, Takada A: Proliferation of donor hematopoietic cells in irradiated and unirradiated host mice. Transplantation 1971;12:344–338.
4 Brecher G, Ansell JD, Micklem HS, Tjio JH, Cronkite EP: Special proliferative sites are not needed for seeding and proliferation of transfused bone marrow cells in normal syngeneic mice. Proc Natl Acad Sci USA 1982;79:5085–5087.
5 Saxe DF, Boggs SS, Boggs DR: Transplantation of chromosomally marked syngeneic marrow cells into mice not subjected to hematopoietic stem cell depletion. Exp Hematol 1984;12:277–283.
6 Stewart FM, Crittenden RB, Lowry PA, Pearson-White S, Quesenberry PJ: Long-term engraftment of normal and post-5-fluorouracil murine marrow into normal nonmyeloablated mice. Blood 1993; 81:2566–2571.
7 Ramshaw HS, Rao SS, Crittenden RB, Peters SO, Weier HU, Quesenberry PJ: Engraftment of bone marrow cells into normal unprepared hosts: Effects of 5-fluorouracil and cell cycle status. Blood 1995;86:924–929.
8 Ramshaw H, Crittenden RB, Dooner M, Peters SO, Rao SS, Quesenberry PJ: High levels of engraftment with a single infusion of bone marrow cells into normal unprepared mice. Biol Blood Marrow Transplant 1995;1:74–80.
9 Rao SS, Peters SO, Crittenden RB, Stewart FM, Ramshaw HS, Quesenberry PJ: Stem cell transplantation in the normal nonmyeloablated host: Relationship between cell dose, schedule and engraftment. Exp Hematol 1997;25:114–121.
10 Kaufman CL, Colson YL, Wren SM, Watkus S, Simmons RL, Ildstad ST: Phenotypic characterization of a novel bone marrow-derived cell that facilitates engraftment of allogeneic bone marrow stem cells. Blood 1994;84:2436–2446.
11 Peters SO, Kittler EL, Ramshaw HS, Quesenberry PJ: Murine marrow cells expanded in culture with IL-3, IL-6, IL-11, and SCF acquire an engraftment defect in normal hosts. Exp Hematol 1995; 23:461–469.

12 Peters SO, Kittler ELW, Ramshaw HS, Quesenberry PJ: Ex vivo expansion of murine marrow cells with interleukin-3, interleukin-6, interleukin-11 and stem cell factor leads to impaired engraftment in irradiated hosts. Blood 1996;87:30–37.

13 Reddy GPV, Tiarks CY, Pang L, Quesenberry PJ: Synchronization and cell cycle analysis of pluripotent hematopoietic progenitor stem cells. Blood 1997;90:2293–2299.

14 Kittler ELW, Peters SO, Crittenden RB, Debatis ME, Ramshaw HS, Stewart FM, Quesenberry PJ: Cytokine facilitated transduction leads to low-level engraftment in nonablated hosts. Blood 1997;90:865–872.

15 Sorrentino BP, Brandt SJ, Bodine D, Gottesman M, Pastan I, Cline A, Nienhois AW: Selection of drug-resistant bone marrow cells in vivo after retroviral transfer of human MDR1. Science 1992;257:99–103.

16 Carter RF, Abrahams-Ogg ACG, Dick JE, Kruth SA, Valli VE, Kamel-Reid S, Dube ID: Autologous transplantation of canine long-term marrow culture cells genetically marked by retroviral vectors. Blood 1992;79:356–364.

17 Stewart FM, Zhong S, Quesenberry PJ: Effects of minimal irradiation (100cGy) on donor and/or recipient animals receiving either delayed (six weeks) or immediate bone marrow transplant (abstract). Blood 1997;90:317b.

18 Barranger JA, Rice E, Sansieri C, Bahnson A, Mohney T, Swaney W, Takiyama N, Dunigan J, Beeler M, Schierer-Fochler S, Ball E: Transfer of the glucocerebrosidase gene to CD34 cells and their autologous transplantation in patients with Gaucher disease. Blood 1997;90:405a.

19 Becker PS, Riel GJ, Debatis ME, Quesenberry PJ, Rice E, Sansieri C, Bahnson A, Mohney T, Swaney N, Takiyama N, Dunigan J, Beeler M, Locot S, Schierer-Fochler S, Ball E, Barranger JA: Transfer of the glucocerebrosidase gene to CD34+ cells results in marking of HPP-CFC. Blood 1997;90:237a.

Peter J, Quesenberry, MD, Director, Cancer Center, University of Massachusetts
Medical Center, University of Massachusetts Cancer Center,
Two BioTech Suite 202, 373 Plantation Street, Worcester, MA 01605 (USA)
Tel. +1 (508) 856 6956, Fax +1 (508) 856 1310, E-Mail Peter.Quesenberry@banyan.ummed.edu

Author Index

Allay, J.A. 143

Banerjee, D. 82, 107
Bank, A. 50
Beauséjour, C. 124
Becker, P.S. 172
Bertino, J.R. 82, 107
Blakley, R.L. 143
Blau, C.A. 162

Capiaux, G. 107

Davis, B.M. 65
Diasio, R.B. 115

Eliopoulos, N. 124
Ercikan-Abali, E.A. 82, 107

Gerson, S.L. 65
Glade Bender, J. 1

Hesdorffer, C. 50

Koç, O.N. 65

Liu-Chen, X. 107

MacKenzie, K.L. 20
May, C. 1
Mineishi, S. 82, 95
Momparler, R.L. 124
Moore, M.A.S. 20

O'Connor, O.A. 107

Pavie, A. 85
Piquard, F. 26

Quesenberry, P.J. 172

Reese, J.S. 65
Rivella, S. 1

Sadelain, M. 1
Sorrentino, B.P. 143

Takebe, N. 107
Tong, Y. 107

Ward M. 50

Zhao, S.C. 82

Subject Index

Adeno-associated virus, dihydrofolate reductase gene transduction 96, 97
Adenovirus, dihydrofolate reductase gene transduction 96, 97
O^6-Alkylguanine DNA alkyltransferase gene, see Methylguanine-DNA-methyltransferase gene
5-Aza-2′-deoxycytidine
  characteristics and cancer chemotherapy 126–128, 136
  metabolism, see Cytidine deaminase

Beige/nude/X-linked immunodeficient mouse, engraftment of transduced human pluripotent stem cells 38–40
O^6-Benzylguanine, see Methylguanine-DNA-methyltransferase gene
Breast cancer, hematopoietic stem cell screening 23, 24

CD34, selection for hematopoietic stem cells 2, 23, 29
Centrifugation transduction, retroviruses 36
Cobblestone area forming cells, cytokine priming for transduction 28–33
Cytidine deaminase
  cytosine nucleoside analog metabolism 124, 129
  gene transduction in chemotherapy protection
    dihydrofolate reductase fusion vector 90, 133–135
    inhibitor reversal of chemoprotection 132, 133
    murine hematopoietic cells 135, 136
    optimization 137
    rationale 128, 129
    retroviral vectors and drug resistance 130–133, 136
  human enzyme characteristics 130
  inhibition in chemotherapy 130
  tissue distribution 129
Cytokine priming, stem cell cycle activation for transduction 28–33, 97, 164
Cytosine arabinoside
  characteristics and cancer chemotherapy 125, 128, 136, 137
  metabolism, see Cytidine deaminase

2′,2′-Difluorodeoxycytidine
  characteristics and cancer chemotherapy 126, 128, 136
  metabolism, see Cytidine deaminase
Dihydrofolate reductase
  advantages over MDR-1 in chemoprotection 98
  gene transduction for chemotherapy protection
    coexpression in resistance augmentation
      thymidine kinase 99, 100
      xanthine guanine phosphoribosyl transferase 101–103
    fusion to other drug resistance proteins 90, 133–135

human application in clinical trials 91
murine hematopoietic precursor
   transduction 86–90, 98
mutant forms of enzyme, properties
   84–86, 98
optimization of transduction 97
rationale 128
retroviral transfer in mice and
   chemoprotection 90, 91
vectors
   adeno-associated virus 96, 97
   adenovirus 96, 97
   retrovirus 96
in vivo selection of transduced cells
   advantages 145
   bicistronic vectors containing reporter
      genes 149, 150, 152, 153
   clinical potential 155, 157, 158
   mouse models 145, 146, 155
   mutant enzymes and drug resistance
      148, 149, 157
   nucleotide transport inhibitors,
      increasing of selective pressure
      146–148, 155
   overview 96, 98, 144, 145
   stem cell selection verification 153–155
methotrexate resistance
   mechanism of action 83, 84
   mutation enhancement 82–86
Dihydropyrimidine dehydrogenase
   circadian variation 117
   5-fluorouracil
      enzyme importance in drug
         pharmacology 116–118
      inhibition in drug therapy 118, 121
      resistance role of enzyme 120, 121
   interindividual variability 117
   tumor expression 118

Fibronectin, enhancement of retrovirus
   transduction 37, 38, 97
FK1012, selection of genetically modified
   cells
   candidate molecules for stem cell
      expansion 166, 167
   FKBP12 constructs 165, 166
   principle 164, 165

proliferative signaling activation
   165, 167, 169
reversibility of proliferation 165, 166,
   169
Flow-through transduction, retroviruses
   36, 37
5-Fluorouracil
   dihydropyrimidine dehydrogenase
      drug resistance role 120, 121
      importance in drug pharmacology
         116–118
      inhibition in drug therapy 118, 121
   marrow engraftment levels, effects 174
   metabolism 115, 116
   resistance mechanisms in tumors
      118–121
   therapeutic applications 115

β-Galactosidase, transduction efficiency
   assay 27
Gibbon ape leukemia virus
   envelope in human cell targeting 4–6, 24,
      25, 52
   receptor, see Glvr-1
β-Globin locus control region, see Locus
   control region
Glvr-1, expression on stem cells 24
Glvr-2, phosphate deprivation and
   upregulation 55
Granulocyte colony-stimulating factor,
   stem cell protection in chemotherapy
   107
Granulocyte-macrophage colony-
   stimulating factor, stem cell protection
   in chemotherapy 107
Green fluorescent protein
   in vitro selection marker 146, 163
   transduction efficiency assay 27, 28

Hematopoietic stem cell
   CD34 marker selection 2, 23, 29
   collection sites and efficiency of gene
      transfer 21–23
   cytokine priming for transduction 28–33,
      97, 164
   diseases treated by gene transfer,
      overview 20, 21, 143, 144, 157, 158

Hematopoietic stem cell (continued)
  early cell tolerance to chemotherapy drugs 163, 164
  ex vivo selection 162
  functional definition 1, 51
  in vivo selection, see In vivo selection
  retroviral transduction efficiency factors
    cell division 6
    overview 3, 34
    reverse transcription 6, 7
    titer of virus 7
    viral envelope 3–6, 24–26
Human immunodeficiency virus, cell division independence for transduction 6, 21

In vivo selection
  applications and rationale 144, 145, 162
  dihydrofolate reductase
    advantages 145
    bicistronic vectors containing reporter genes 149, 150, 152, 153
    clinical potential 155, 157, 158
    mouse models 145, 146, 155
    mutant enzymes and drug resistance 148, 149, 157
    nucleotide transport inhibitors, increasing of selective pressure 146–148, 155
    overview 96, 98, 144, 145
    stem cell selection verification 153–155
  FK1012 for selection of genetically modified cells
    candidate molecules for stem cell expansion 166, 167
    FKBP12 constructs 165, 166
    principle 164, 165
    proliferative signaling activation 165, 167, 169
    reversibility of proliferation 165, 166, 169
  genes for selection, types 144, 145, 163
  methylguanine-DNA-methyltransferase 73, 75–77
Insulator elements, rationale for inclusion in retroviral vectors 12
Interleukin-3, stem cell protection in chemotherapy 107

Locus control region
  β-globin locus control region, fragment inclusion in retroviral vectors 11
  incorporation and effect on viral titer 7
  rationale for vector inclusion 9–11
Long-term culture-initiating cells
  cytokine priming for transduction 28–33
  identification 2
  retroviral transduction efficiency 5, 6
Long terminal repeat, driving of transgene expression 8, 9

Matrix attachment region
  apolipoprotein B gene 12
  interferon-γ 12
  rationale for inclusion in retroviral vectors 11, 12
Methotrexate
  chemotherapy toxicity protection, see Dihydrofolate reductase
  leucovorin treatment of toxicity 82
  therapeutic applications 82
Methylguanine-DNA-methyltransferase gene
  O^6-benzylguanine-resistant mutants
    dose response of cytoprotection in mice 69, 70
    engineering from bacterial protein 70
    G156A mutant transduction cytoprotection 71–73
    in vivo selection for transduced cells 73, 75–77
    preclinical xenograft studies 73
    selection for transduced cells in nonmyeloablated mice 76, 77
  mechanism of wild-type enzyme inactivation 68, 69
  DNA repair mechanism 65–67
  expression levels in hematopoietic cells 65, 67
  mutagen protection 65, 66
  wild-type gene transfer, limitations 67, 68, 77
Mononuclear cell, cytokine priming for gene transfer 23

Subject Index

182

Multidrug resistance gene, transfer in
    chemotherapy 57
Multiple drug resistance gene
    murine cells in optimization of transfer to
        human progenitor cells
        fetal liver cells in assay 54, 55
        marrow cells 52–54
    retroviral transduction 51, 52
    transfer in chemotherapy
        cell selection and transplantation in
            bone marrow 51
        clinical trials 57–59
        growth factor stimulation 58, 60
        hematopoietic stem cell limitations
            59, 60
        human marrow transduction 56
        limitations 50, 51
        peripheral blood progenitor cell
            transduction 56, 57
        rationale 50, 128
Murine leukemia virus
    amphotropic virus
        receptor, see Glvr-2, Ram-1
        transduction efficiency 4
    cell division requirement for transduction
        6, 13
    envelopes in human cell targeting 4, 5,
        24, 25
    long terminal repeat-driven transgene
        expression 8, 9
    packaging cell lines 4, 52
Myeloablation, inhibition of gene therapy
    172, 176, 177

Neomycin resistance gene, transduction
    efficiency assays 26, 27, 53, 88, 96
Nerve growth factor receptor, transduction
    efficiency assay 27
Nitrobenzylmercaptopurine riboside,
    sensitization to trimetrexate for in vivo
    selection 147, 148, 155
NOD/SCID mouse, engraftment of
    transduced human pluripotent stem cells
    38, 40, 41
Nonmyeloablated hosts
    engraftment levels
        cytokine effects 174

5-fluorouracil effects 174
MDR1-transduced cells 175
mice 172, 173
gene therapy success
    donor to host ratio of differentiated
        marrow and blood cells 175, 176
    stem cell strategy 176, 177
gene transfer into dogs 175

Peripheral blood progenitor cell
    bone marrow transplantation 51
    MDR transduction 56, 57, 61
Phosphate deprivation, retroviral receptor
    upregulation 55
Polymerase chain reaction, transduction
    efficiency assays 26, 53, 58, 59, 88, 89

Ram-1
    expression on stem cells 24, 55
    phosphate deprivation and upregulation
        55
Retroviral transduction strategies
    centrifugation transduction 36
    fibronectin enhancement 37, 38, 97
    flow-through transduction 36, 37
    producer cell coculture 35, 36
    static supernatant transduction 35
Reverse transcription, efficiency in gene
    transduction 6, 7

Scaffold attachment region, see Matrix
    attachment region
SCID-hu mouse, engraftment of transduced
    human pluripotent stem cells 39
Spinoculation, retrovirus transduction 36

Thymidine kinase
    coexpression with dihydrofolate reductase
        in chemoprotection 99, 100
    suicide gene eradication of tumor
        cells 100, 101
Thymidylate synthase
    chemotherapy targeting drugs 108
    5-fluorouracil resistance role 119
    gene transduction for chemotherapy
        protection
        dihydrofolate reductase gene fusion 90

Thymidylate synthase, chemotherapy
  targeting drugs (continued)
    engineering of human mutants
      characterization of mutants 110, 111
      conserved regions 109
      random ethylmethane sulfonate
        mutagenesis 109, 110
      site-directed mutagenesis 110
      Y33H 108, 109
    human studies 112
    retroviral transfer and chemoprotection
      111, 112
    reaction catalyzed 108
Transforming growth factor-β, inhibition of
  retroviral transduction 33, 34

Trimetrexate toxicity, *see* Dihydrofolate
  reductase

Vesicular stomatitis virus, G protein
  envelope
    human cell targeting 4, 5, 24,
      25, 52
    toxicity 25, 52

Xanthine guanine phosphoribosyl
  transferase
    coexpression with dihydrofolate reductase
      in chemoprotection 101–103
    suicide gene eradication of tumor
      cells 103